DISTRIBUTED RENEWABLE ENERGIES FOR OFF-GRID COMMUNITIES

Strategies and Technologies toward Achieving Sustainability in Energy Generation and Supply

DISTRIBUTED RENEWABLE ENERGIES FOR OFF-GRID COMMUNITIES

Strategies and Technologies toward Achieving Sustainability in Energy Generation and Supply

NASIR EL BASSAM

PREBEN MAEGAARD

MARCIA LAWTON SCHLICHTING

ELSEVIER

Amsterdam • Boston • Heidelberg • London • New York • Oxford
Paris • San Diego • San Francisco • Singapore • Sydney • Tokyo

Elsevier
The Boulevard, Langford Lane, Kidlington, Oxford, OX5 1GB, UK
225 Wyman Street, Waltham, MA 02451, USA

First published 2013

Notices

Knowledge and best practice in this field are constantly changing. As new research and experience broaden our understanding, changes in research methods, professional practices, or medical treatment may become necessary.

Practitioners and researchers must always rely on their own experience and knowledge in evaluating and using any information, methods, compounds, or experiments described herein. In using such information or methods they should be mindful of their own safety and the safety of others, including parties for whom they have a professional responsibility.

To the fullest extent of the law, neither the Publisher nor the authors, contributors, or editors, assume any liability for any injury and/or damage to persons or property as a matter of products liability, negligence or otherwise, or from any use or operation of any methods, products, instructions, or ideas contained in the material herein.

Library of Congress Cataloging-in-Publication Data
El Bassam, Nasir
   Distributed renewable energies for off-grid communities : planning, technologies, and applications / N. El Bassam, P. Maegaard, Marcia Lawton Schlichting.
      p. cm.
   Includes bibliographical references.
   ISBN 978-0-12-397178-4
1. Small power production facilities. 2. Distributed generation of electric power. 3. Renewable energy sources. 4. Electric power distribution. 5. Energy development. I. Maegaard, Preben. II. Schlichting, Marcia Lawton. III. Title.
   TK1006.E43 2013
   333.79'4—dc23
                                                    2012028410

British Library Cataloguing-in-Publication Data
A catalogue record for this book is available from the British Library.

ISBN: 978-0-12-397178-4

For information on all Elsevier publications visit our
website at http://store.elsevier.com

Working together to grow
libraries in developing countries

www.elsevier.com | www.bookaid.org | www.sabre.org

ELSEVIER   BOOK AID
            International   Sabre Foundation

CONTENTS

PREFACE

The time of cheap oil and gas is over. Mankind can survive without globalization, financial crises and flights to the moon or Mars but not without adequate and affordable energy availability.

Energy is directly related to the most critical economic and social issues that affect sustainable development such as water supply sanitation, mobility, food production, environmental quality, education, job creation, security and peace in regional and global contexts. Indeed the magnitude of change needed is immense, fundamental and directly related to the energy produced and consumed nationally and internationally. In addition, it is estimated that almost two billion people worldwide lack access to modern energy resources.

Current approaches to energy are non-sustainable and non-renewable. Today, the world's energy supply is largely based on fossil fuels and nuclear power. These sources of energy will not last forever and have proven to be contributors to our environmental problems. In less than three centuries since the industrial revolution, mankind has already burned roughly half of the fossil fuels that accumulated under the earth's surface over hundreds of millions of years. Nuclear power is also based on a limited resource (uranium) and the use of nuclear power creates such incalculable risks that nuclear power plants cannot be insured. After 50 years of intensive research, no single safe long-term disposal site for radioactive waste has been found.

Although some of the fossil energy resources might last a little longer than predicted, especially if additional reserves are discovered, the main problem of "scarcity" will remain, and this represents the greatest challenge to humanity.

Renewable energy offers our planet a chance to reduce carbon emissions, clean the air, and put our civilization on a more sustainable footing. Renewable sources of energy are an essential part of an overall strategy of sustainable development. They help reduce dependence of energy imports, thereby ensuring a sustainable supply and climate protection. Furthermore, renewable energy sources can help improve the competitiveness of industries over the long run and have a positive impact on regional development and employment. Renewable energies will provide a more diversified, balanced, and stable pool of energy sources.

Some countries of the EU such as Denmark, China, Germany, Austria and Spain as well as China and India have already demonstrated the impressive pace of transition that can be achieved in renewable energy deployment, if the right policies and frameworks are in place. Also the new US policy has made clear its determination to massively increase renewable energy in the US, giving strong and clear signals to the world.

The main target of this book will be a comprehensive and solid contribution to enlighten the vital role of developing decentralized and distributed renewable energy production and supply for off-grid communities along with their technical feasibilities to meet the growing demand for energy and to face the present and future challenges of limited fossil and nuclear fuel reserves, global climate change and financial crises. It deals also with various options and case studies related to the potential of renewable energies and the future transition versions along with their environmental, economic and social dimensions.

With rapid and continued growth in the world, it is no longer a question of when we will incorporate various renewable energy sources into the mix, but how fast the transition can be managed.

The Authors
Prof. Dr. N. El Bassam
Prof. P. Meagaard
M. Schlichting
June 2012

LIST OF FIGURES

Figure 1.1: Relevance of distributed generation.
India Energy Portal, Distributed Generation, www.indiaenergyportal.org/ subthemes.php?text=

Figure 1.2a, b: Stand-alone off-grid systems.
http://www.wholesalesolar.com/products.folder/systems-folder/ OFFGRID

Figure 1.3: Off-grid system which can be also connected to the grid.
Reprinted with permission. ©2012 Home Power Inc., www.homepower. com

Figure 1.4: Combined power plant.
http://www.blog.thesietch.org/2007/12/30/germany-going-100-renewable-or-yet-another-reason-why-america-is-falling-behind/
Accessed March 23, 2010

Figure 2.1: Past, present and future energy sources.
Source: El Bassam, 1992

Figure 2.2: Global final energy consumption, 2006 (REN21, 2007).
Source: REN21 Renewables 2007 Global Status Report, Copyright © 2008 Deutsche Gesellschaft für Technische Zusammenarbeit (GTZ) GmbH. www.ren21.net

Figure 2.3: Regional breakdown of world energy demand scenario in the new policies.
Source: Projection of EIA 2009

Figure 2.4: Peak oil. The decline of global production output will lead to peak cheap oil by 2012 (Projection of EIA 2009).

Figure 2.5: Global energy demand and resources.
Source: BSU Solar, BSU Solar (German Solar Industry Assoc.), 2007

Figure 2.6: Physical potential of renewable energies.
Source: Nitsch, F., BMU documentation, 2007

Figure 3.1: Sustainability in regional and global context demands, risks and measures.
Source: El Bassam, 2004

Photo: Georg Slickers, 2006 (http://en.wikipedia.org/wiki/Creative_ Commons)

Content Last Updated. 02/09/2011

Source: German American Bioenergy Conference / Syracuse, Eckhard Fangmeier 2009-06-23

Figure 14.4: Wildpoldsried – The 100% renewable energy town.
Photo credit: City of Wildpoldsried accessed from Internet 2012, http://www.go100percent.org/cms/index.php?id=19&id=69&tx_ttnews[tt_news]—111&tx_locator_pi1[startLat]=45.93583305&tx_locator_pi1[startLon]=-5.6139353&cHash=b2670e5ad38be35e10c09dd0a257a7e8

Figure 14.5a: Energy sources used by IES Pty Ltd to provide essential services to Indigenous communities in the NT, and b: Map details.
(Northern Territory Government Australia: Roadmap to Renewable and Low Emission Energy in Remote Communities Report (pdf download), 2011)

Figure 14.6: Schematic of conventional power station vs. stabilized renewable energy power station.
(Northern Territory Government Australia: Roadmap to Renewable and Low Emission Energy in Remote Communities Report (pdf download), 2011)

Figure 14.7: Graphic illustration of a "sustainable community" within Iraq.
(Iraq Dream Homes, LLC (IDH) 2010, http://www.iraq-homes.com/about_us-en.html)

Figure 14.8: Solar panels power street lights in Fallujah, Iraq.
Credit: U.S Army 2010, http://www.army.mil/article/32799/

Figures 14.9–14.33 are accessed from: Maegaard, P., Thisted, 100% Renewable Energy municipality. PowerPoint presentation, presented 2009/2010 at conferences/events in 34 cities in 18 countries.

Figure 14.9: The electrical system.

Figure 14.10: (left) Thisted combined heat and power for town of 25,000 residents. Fuels used include household waste, straw, wood and geothermal. (center) 6 MW_{el} steam turbine; (right) trucks discharge waste.

Figure 14.11: Denmark power stations.

Figure 14.12: Biomass from the agriculture and forests is a limited resource. Dry biomass is ideal for long-term and seasonal storage of energy when solar and wind energy is not available.

Figures 15.1 Renewable ownership; 15.2 Wind ownership; and 15.3 Solar energy ownership: Paul Gipe, Chart by Paul Gipe from data by Unendlich viel energie, Windworks.org, January 5, 2012. http://wind-works.org/coopwind/CitizenPowerConferencetobeheldinHistoricChamber.html

Figure 15.4: Global new investments in renewable energy.
Data source: Bloomberg New Energy Finance, UNEP SEFI, Frankfurt School, Global Trends in Renewable Energy Investment.

Figure 15.5: Clean energy projected growth 2007-2017.
Based on Clean Edge (2008) Energy Trends Report (GGByte at en.wikipedia 2012), http://en.wikipedia.org/wiki/File:Re_investment_2007-2017.jpg

Figures A5.1a–A5.2 accessed from Emna Latiri, 2012, architect, personal communication, www.solartech-sud.com/eng/.

Figure A5.1a: Virtual image of the 'Olive Branch', A5.1b: Solar Eco-Village Zarzis, A5.1c: Map of Tunisia.
(Emna LATIRI, Architect 2012)

Figure A5.2: Virtual image of the Solar-Eco village.
(Emna LATIRI, Architect 2012)

Figure A6.1: Aerial photo of the Solar Park Wierthe, Germany.

Figure A6.2 and A6.3: A boy (Janosch Goosse) is expecting a sustainable and safe power supply for his future.
(El Bassam, 2012)

Figure A6.4: (a) First solar desalination unit and (b) electric car are demonstrated at IFEED-Research Centre and Solar Park Vechelde.
(El Bassam, 2012)

Figure A7.1: Eternal University, Baru Sahib, India.
(self-photos, Eternal University, Baru Sahib, Via Rajgarh, Sirmour, Himachal Pradesh, India – 173101)

Figure A7.2: The valley of Baru Sahib, Eternal University India with solar heating and power supply systems.
(self-photos, Eternal University)

Figure A7.3: Solar devices on the buildings and valley slopes.
(self-photos, Eternal University)

Fig A7.4: Laundry building supplied with electricity (PV) and heated water (solar thermal).
(self-photos, Eternal University)

Figure A7.5: Laundry building.
(self-photos, Eternal University)

Figure A8.1 (a) With external electrical pump, (b) Clean water from very dirty water.
(Best Water Source, Ernst Hauseder – Austria, hauseder@ettl.net)

LIST OF TABLES

CHAPTER ONE

Scope of the Book

Distribution means the delivery of electricity to the retail customer's home or business through low-voltage distribution lines. Distributed generation is also called on-site generation, dispersed generation, embedded generation, or decentralized generation. Decentralized energy, or distributed energy, generates electricity, heat and fuels from many small energy sources. It reduces the amount of energy lost in transmitting electricity because the electricity is generated very near where it is used, perhaps even in the same building. This also reduces the size and number of power lines that must be constructed. Both electric demand reduction (energy conservation, load management, etc.) and supply are generated at or near where the power is used. A distributed generation system involves amounts of generation located on a utility's distribution system for the purpose of meeting local (substation level) peak loads and/or displacing the need to build additional (or upgrade) local distribution lines.

1.1. DISTRIBUTED ENERGY GENERATION

Distributed generation is also defined as installation and operation of small modular power-generating technologies that can be combined with energy management and storage systems. It is used to improve the operations of the electricity delivery systems at or near the end user. These systems may or may not be connected to the electric grid.

1.2. DISTRIBUTED ENERGY SUPPLY

Typical distributed power sources in a Feed-in Tariff (FIT) scheme have low maintenance, low pollution and high efficiencies. In the past, these traits required dedicated operating engineers and large complex plants to reduce pollution. However, modern embedded systems can provide these traits with automated operation and renewables, such as sunlight, wind and geothermal. This reduces the size of a profitable power plant.

In the future, when planning a new power and heating or transportation fuels system on a clean sheet of paper, there will not be any big fossil-fuel

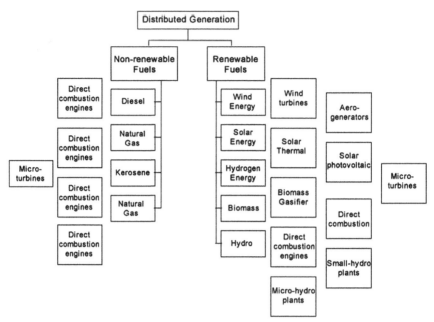

Figure 1.1 *Relevance of distributed generation. (India Energy Portal, Distributed Generation). www.indiaenergyportal.org/subthemes.php?text=.*

based power stations or big, high-voltage transmission lines. Each community in this new energy supply structure will have a variety of local supply technologies based on solar, wind, biomass and other locally available sources of energy.

Solar and wind will be the primary sources of supply. Biomass will be especially important for complementary power and heat when solar and wind is not sufficient. Balancing of fluctuating power from solar and wind is necessary; chemical and thermal storage solutions can also be applied. Such future supply systems can be of many sizes. The smallest will be for one family house or settlement and the biggest for a region or city.

As small-scale technologies can be mass produced and are therefore cheap, it may well prove preferable and most economical to divide up cities into many independent and autonomous decentralized systems. In this book we will call them *off-grid*, as there is no advantage to having international or interregional grid structures. Up to this point, off-grid supply has referred to unserved areas in developing countries without a national grid.

Distributed energy resource systems are small-scale power generation technologies (typically in the range of 3 kW to 10,000 kW) used to provide an alternative to or an enhancement of the traditional electric power system.

As the cost of fossil fuels increases and the cost of renewable energy technologies declines and government support increases, investors, utilities, and governments are exploring ways to implement or invest in distributed renewable energy programs, projects and companies.

1.3. COMMUNITY POWER

Community is a term that has different meanings for different people. In this book a community is defined as a social group of any size whose members live in a specific place. The term thus relates to geographical proximity, or "communities of locality" (Walker 2008), such as a neighborhood, town, district or city. This book focuses on distributed energy that is generated and distributed to consumers within a geographic locality. Distributed energy generation can be a continuum of energy generation from a household and multiple-buildings scale to a larger-community scale. Some energy may be fed back into the electricity grid, but ideally at least some of the total energy generated is distributed and consumed locally.

The book profiles the special topic of Community Power – Citizens' Power, referring to the development and ownership of renewable energy projects by local citizens and communities, including farmers and land-owners, cooperatives, municipalities, local and regional developers and utilities.

1.4. OFF-GRID SYSTEMS

Off-grid systems provide an independently regulated power supply that has at least the same reliability and quality as a public power grid.

The term off-grid refers to not being connected to a grid, mainly used in terms of not being connected to the main or national transmission grid in electricity. Off-grid electrification is an approach to access electricity used in countries and areas with little access to electricity, due to scattered or distant population. It can be any kind of electricity generation. Electrical power can be generated on-site with renewable energy sources such as solar, wind or geothermal; or with a generator and adequate fuel reserves.

It can also connect to local and national grids to substitute for energy supply generated by nuclear or other non-renewable fuel sources, which is called *green electricity* in some industrialized countries.

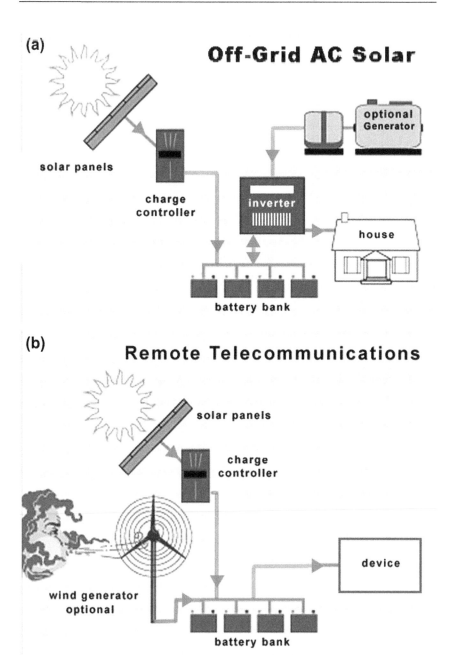

Figure 1.2a, b *Stand-alone off-grid systems. http://www.wholesalesolar.com/products. folder/systems-folder/OFFGRID.*

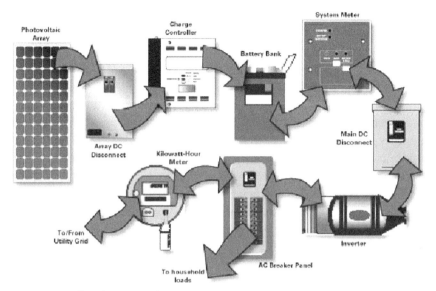

Figure 1.3 *Off-grid system which can also be connected to the grid.* Reprinted with permission. ©2012 Home Power Inc., www.homepower.com.

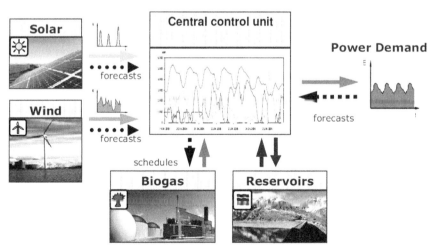

Figure 1.4 *Combined power plant.* http://www.blog.thesietch.org/2007/12/30/germany-going-100-renewable-or-yet-another-reason-why-america-is-falling-behind/. (Accessed March 23, 2010). (See color plate 1.)

This book illustrates the future of off grid power supply, as there is no advantage to having international and/or interregional grid structures. So far, *off-grid supply* has generally referred to communities in developing countries without a national grid. Due to the decentralized character of the renewable energies, the future rationale will be to apply off-grid technologies to all types of communities, urban and rural, in fossil-fuel served areas in the industrialized countries, and for unserved regions in developing countries as well. What thus far has been the exception may, due to the transition to decentralized energy forms, change to become mainstream.

Similar fundamental changes have often appeared when considering the long historical perspective. Facing the end of the fossil-fuel age and the enormous risks of the ongoing climate change, it is time to prepare for the exit of the fossil-fuel era. And there are no technological barriers—this is the important message of the book.

REFERENCES

Gangwar, R., 2009. Building community resilience towards climate change adaptation through awareness and education. Paper presented at the seminar Energy and Climate in Cold Regions of Asia, 21–24. April, 2009, available at. http://india.geres.eu/docs/Seminar_proceedings/3-Climate_Change_Impacts_and_Adaptation/Building%20Community%20Resilience%20towards%20Climate%20Change%20Adaptation%20through%20Awareness%20&%20Education.pdf.

Renewable Communities, 2008. Renewable communities: Moving towards community resiliency [online] available at. http://renewablecommunities.wordpress.com.

CHAPTER TWO

Restructuring Future Energy Generation and Supply

2.1. BASIC CHALLENGES

By 2050, humanity will need two to three earths to cover its consumption of resources, if we continue to manage our resources as business as usual.

The global energy system currently relies mainly on hydrocarbons such as oil, gas and coal, which together provide nearly 80 percent of energy resources. Traditional biomass—such as wood and dung—accounts for 11 percent and nuclear for 6 percent, while all renewable sources combined contribute just 3 percent. Energy resources, with the exception of nuclear, are ultimately derived from the sun. Non-renewable resources such as coal, oil and gas are the result of a process that takes millions of years to convert sunlight into hydrocarbons. Renewable energy sources convert solar radiation, the rotation of the earth and geothermal energy into usable energy in a far shorter time.

In the International Energy Agency (IEA) Reference Scenario, world primary energy demand grows by 1.6% per year on average in 2006–2030, from 472 exajoules (EJ) (11,730 Mtoe—million metric tons of oil equivalents) to just over 714 EJ (17,010 Mtoe). Due to continuing strong economic growth, China and India account for just over half of the increase in world primary energy demand between 2006 and 2030. Middle East countries strengthen their position as an important demand center, contributing a further 11% to incremental world demand. Collectively, non-OECD (Organization for Economic Cooperation and Development) countries account for 87% of the increase. As a result, their share of world primary energy demand rises from 51% to 62%. Their energy consumption overtook that of the OECD in 2005 (OECD/IEA 2008).

The major challenges of today's energy system are closely related to a wide scale of essentials such as welfare, dignity, peace, nature and sustainable development. They include:

- Limited oil, gas and uranium resources.
- Almost 2 billion of the world's population have no access to electricity, gas, oil or clean water.

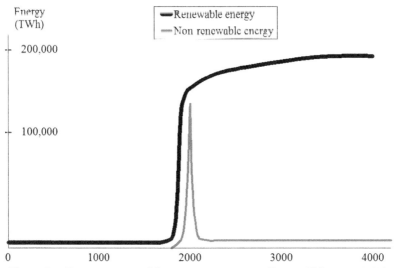

Figure 2.1 *Past, present and future energy sources. (Source: El Bassam, 1992).*

- Increasing import dependency in most industrialized countries, China and India.
- Energy prices and volatility (the time of cheap oil and gas is over!).
- Climate change and other environmental risks (energy accounts for 80% of all greenhouse gas (GHG) emission).
- Geo-strategic tensions caused by scarce energy resources.
- The extracting, transport, processing and use of fossil and nuclear fuels can be eventually threatening to nature and existence of mankind (i.e., accidents in Gulf of Mexico and Fukushima).
- Worldwide, there is no single safe long-term storage facility for highly radioactive nuclear wastes.
- Growing world population (2012: 7 billion, 2050: 9–10 billion).

Figure 2.1 effectively demonstrates the finiteness of fossil energy resources and the vital role of renewable energies in satisfying the needs of the present and future generations for adequate and affordable energy to ensure sustainable development.

2.2. CURRENT ENERGY SUPPLIES

Current global energy final supplies are dominated by fossil fuels (388 EJ per year), with much smaller contributions from nuclear power (26 EJ)

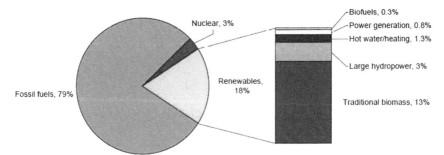

Figure 2.2 *Global final energy consumption, 2006 (REN21, 2007). (Source: REN21 Renewables 2007 Global Status Report).* (See color plate 2.)

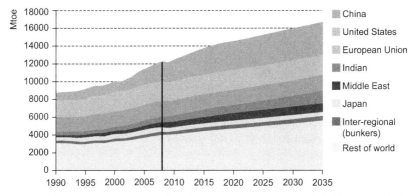

Figure 2.3 *Regional breakdown of world energy demand scenario in the new policies. (Source: Projection of EIA 2009).* (See color plate 3.)

and hydropower (28 EJ). Biomass provides about 45 ± 10 EJ, making it by far the most important renewable energy source used.

Renewable energy supplies 18 percent of the world's final energy consumption, counting traditional biomass, large hydropower, and "new" renewable (small hydro, modern biomass, wind, solar, geothermal, and biofuels) (Figure 2.2). Traditional biomass, primarily for cooking and heating, represents about 13 percent and is growing slowly or even declining in some regions as biomass is used more efficiently or replaced by more modern energy forms.

2.3. PEAK OIL

Peak oil theory states that any finite resource (including oil) will have a beginning, middle, and an end of production, and at some point it will reach a level of maximum output. Today we consume around four times as much oil as we discover.

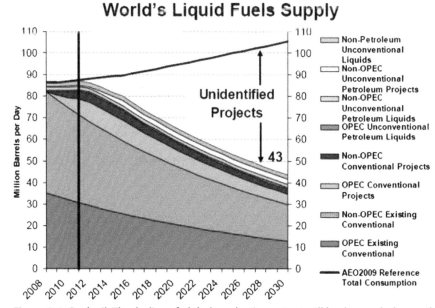

Figure 2.4 *Peak oil: The decline of global production output will lead to peak cheap oil by 2012. (Projection of EIA 2009).* (See color plate 4.)

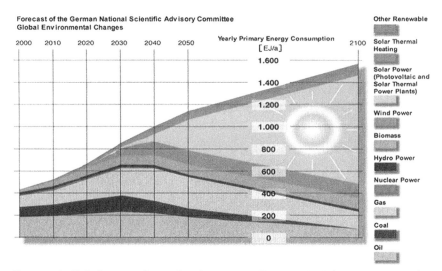

Figure 2.5 *Global energy demand and resources. (Source: BSU Solar, 2007).* (See color plate 5.)

The peaking of oil extraction was already reached in:

- USA: 1974
- Venezuela: 1998
- Norway, Oman: 2001
- Mexico: 2004
- Nigeria: 2005

The "Unidentified Projects" in Figure 2.4 refers to the potential of renewable energy solutions that require immediate actions, which should be at the top of the agenda.

The possible future contribution of various renewable energy technologies, which should remain the ultimate priority (solar, wind, hydro, biomass, geothermal...) is demonstrated in Figure 2.5.

2.4. AVAILABILITY OF ALTERNATIVE RESOURCES

Energy cannot be created; it can be converted from one form to other by technical, biological and chemical means, such as solar and wind energy into heat and power energy, biomass into heat, electricity or biofuels, etc. The good news is:

We have all we need of energy resources as well as conversion technologies in order to ensure a complete supply of clean and green energy!

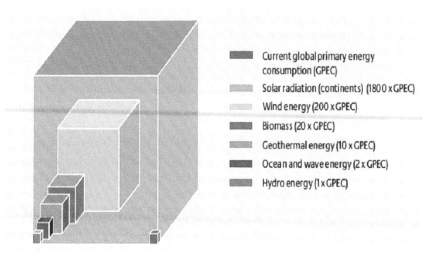

Current global primary energy consumption (GPEC)

Solar radiation (continents) (1800 x GPEC)

Wind energy (200 x GPEC)

Biomass (20 x GPEC)

Geothermal energy (10 x GPEC)

Ocean and wave energy (2 x GPEC)

Hydro energy (1 x GPEC)

Figure 2.6 *Physical potential of renewable energies.* *(Source: Nitsch, F. 2007).* (See color plate 6.)

Renewable energy offers our planet a chance to reduce carbon emissions, clean the air, and put our civilization on a more sustainable footing. Renewable sources of energy are an essential part of an overall strategy of sustainable development. They help reduce dependence on energy imports, thereby ensuring a sustainable supply and climate protection. Furthermore, renewable energy sources can help improve the competitiveness of industries over the long run and have a positive impact on regional development and employment. Renewable energies will provide a more diversified, balanced, and stable pool of energy sources. With rapid growth in Brazil, China and India, and continued growth in the rest of the world, it is no longer a question of when we will incorporate various renewable energy sources more aggressively into the mix, but how fast?

The technically exploitable amounts of energy in the form of electricity, heat and chemical energy from renewable sources exceed the current world energy consumption by about sixfold (Nitsch 2007).

The transition into distributed and decentralized renewable energy systems has to be associated with multiple measures:

- Renewable energies should remain the priority (solar, wind, hydro, biomass, geothermal…)
- Improving the energy efficiency
- Construction of smart grids
- Creating power storage facilities
- Future-oriented and innovative policy in national, regional and global contexts
- Creating a global climate framework
- Intensifying research and education activities
- Improving the cooperation between nations as well as between public and private sectors.

Renewable energy and energy efficiency do not represent an alternative to fossil resources; they are the only options that can ensure sustainable development and the survival of mankind and the protection of our climate.

The share of renewable energy in total energy supply needs to grow by 2% per year in order to ensure future energy demand and to avoid regional and global crises. Although more than 100 years' supply of crude oil is left in the ground, the resources that are "cheap and easy" to extract have for the most part already been discovered and adequate proportions of gas and oil resources should be reserved for human welfare in decades to come.

REFERENCES

BSU Solar, 2007. German Solar Industry Association.

Nitsch, F., 2007. BMU documentation.

Peak Oil news & message boards, 2011. Exploring Hydrocarbon Depletion [online], available at. http://peakoil.com/what-is-peak-oil/.

REN21 Renewables, 2007. Global Status Report, 2008 Deutsche Gesellschaft für Technische Zusammenarbeit (GTZ) GmbH. www.ren21.net.

WEC Global Issues Map, 2011. www.worldenergy.org/wec_news/press_releases/3264.asp.

CHAPTER THREE

Road Map of Distributed Renewable Energy Communities

3.1. ENERGY AND SUSTAINABLE DEVELOPMENT

The world still continues to seek energy to satisfy its needs while giving due consideration to the social, environmental, economic and security impacts. It is now clear that current approaches to energy are unsustainable. It is the responsibility of political institutions to ensure that technologies that enable sustainable development are transferred to the end users. Scientists and individuals bear the responsibility of understanding the earth as an integrated whole and must recognize the impact of our actions on the global environment, in order to ensure sustainability and avoid disorder in the natural life cycle. Wise policy in a regional and global context requires that demands are to be satisfied and risks be avoided.

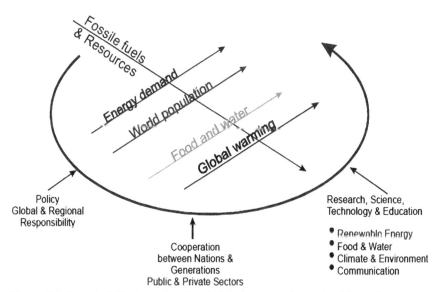

Figure 3.1 *Sustainability in regional and global context demands, risks and measures. (El Bassam 2004).*

Current approaches to energy are unsustainable and non-renewable. Furthermore, energy is directly related to the most critical social issues affecting sustainable development: poverty, jobs, income levels, access to social services, gender disparity, population growth, agricultural production, climate change, environmental quality and economic/security issues. Without adequate attention to the critical importance of energy to all these aspects, the global social, economic and environmental goals of sustainability cannot be achieved. Indeed, the magnitude of change needed is immense, fundamental and directly related to the energy produced and consumed nationally and internationally. The key challenge to realizing these targets is to overcome the lack of commitment and to develop the political will to protect people and the natural resource base. Failure to take action will lead to continuing degradation of natural resources, increasing conflicts over scarce resources and widening gaps between rich and poor. We must act while we still have choices. Implementing sustainable energy strategies is one of the most important levers humankind has for creating a sustainable world. More than two billion people, mostly living in rural areas, have no access to modern energy sources. Food and fodder availability is very closely related to energy availability.

3.2. COMMUNITY INVOLVEMENT

There are different views on what scale of community involvement is advisable or even possible, and a range of models for how distributed renewable energy projects with some form of community involvement can be designed. This book includes a range of case studies with different renewable energy types, geographies, ownership models and types of community participation. But for all case studies selected, communities are involved in some ongoing way with the project, even if local people were not the initiators of the renewable energy project. Communities also have a powerful role to play in rapidly reducing greenhouse gas emissions across the globe. Many communities are already leading the response to climate change by establishing renewable energy production in their town or city, at the same time strengthening relationships among community members, creating local jobs and often generating income for the community.

Community energy includes the social process of establishing and distributing renewable energy technology locally, with social and economic benefits to that defined community. Community energy is therefore about the social arrangements around how an important technology that contributes to the sustainability solution is implemented and brings benefit to people.

The localized distribution of energy has considerable advantages for efficiency and sustainability. But how communities are involved in the initiation, development and consumption of locally produced energy is also important. If energy is produced on a local scale but does not involve or benefit local people, arguably the full sustainability benefits of "community energy" will not be achieved. In terms of achieved renewable energy outcomes, research also suggests that there is a lesser acceptance of renewable energy projects if local people are not involved and benefits are not shared among community members.

Most energy generation is centralized, involving the production of electricity at a large, central facility, the transmission of high-voltage electricity over long distances through the power grid, and the conversion and distribution of that energy to a large number of consumers. Energy may travel many hundreds of kilometers from where it was produced before it reaches the user of the energy. Most centralized production facilities use non-renewable sources such as coal, oil or nuclear material to power their electrical generation. Significant wastage occurs during the transmission of high-voltage electricity over large distances: losses through the grid can amount to 7% to 15% of generation electricity (IEA 2005).

As energy markets are restructured, customers and utilities feel more pressure to control costs and increase operating flexibility. Contributing to this trend is a heightened concern about energy security and the emergence and advanced development of modular renewable generation technologies. The environmental benefits of these distributed power sources exploiting, for example, renewable resources or combined heat and power are substantial.

3.3. FACING THE CHALLENGES

Some appealing concepts for fostering rural development imply the systematic exploitation of locally available renewable sources of energy. In this perspective, Integrated Energy Farms, or Integrated Energy Settlements, can be thought of as sustainable power centers, supplying local markets comfortably with electricity and fuels, while at the same time covering their demand for food crops and soft commodities. Different ecological and socio-cultural settings require specific types of design for locally adapted Integrated Energy Communities.

In order to meet challenges, future energy policies should put more emphasis on developing the potential of energy sources, which should form the foundation of future global energy structure. In this context, the Food and Agriculture Organization (FAO) of the United Nations and the Sustainable Rural Energy Network (SREN) have developed the concept of optimization, evaluation and implementation of integrated renewable energy farms in rural communities (El Bassam 1999).

3.4. THE CONCEPT OF FAO, UN INTEGRATED ENERGY COMMUNITIES (IEC)

Renewable energy has the potential to bring power to communities, not only in the literal sense, but by transforming their development prospects. There is tremendous latent demand for small-scale, low-cost, off-grid solutions to people's varying energy requirements.

People in developing countries understand this only too well. If they were offered new options that would truly meet their needs and engage them in identifying and planning their own provision, then success in providing renewable energy services could become a reality.

Energy is one of the important inputs to empower people, provided it is made available to the people in unserved areas on an equitable basis. Therefore, access to energy should be treated as a fundamental right for everyone. This is only possible if the end users are made the primary stakeholders in the production, operation and management of the generation of useful energy.

Despite much well-intended effort, little progress has been made and a radical new approach is called for, based on the following imperatives:

- Rural development in general and rural energy development specifically need to be given higher priority by policy makers.
- Rural energy development must be decentralized and local resources managed by rural people.
- Rural energy development must be integrated with other aspects of rural development, overcoming the institutional barriers between agriculture infrastructure and education as well as in the social and political spheres.
- Due to the problem of non-committing policies on the government level, different priorities in each country should be addressed.

The productivity and health of a third of humanity are diminished by a reliance on traditional fuels and technologies, with women and children suffering most (El Bassam 1999). Current methods of energy production, distribution and use worldwide are major contributors to environmental problems including global warming and ecosystem degradation at the local, regional and global levels.

The IEC concept includes farms or decentralized living areas from which the daily necessities (water, food and energy) can be produced directly on-site with minimal external energy inputs. Energy production and consumption at the IEC has to be environmentally friendly, sustainable and ultimately based mainly on renewable energy sources. It includes a combination of different possibilities for non-polluting energy production, such as modern wind and solar electricity production, as well as the production of energy from biomass. It should seek to optimize energetic autonomy and an ecologically semi-closed system, while also providing socioeconomic viability and giving due consideration to the newest concepts of landscape and bio-diversity management.

3.5. GLOBAL APPROACH

In order to meet challenges, future energy policies should put more emphasis on developing the potential of energy sources, which should form the foundation of future global energy structure. In this context, the FAO of the United Nations in support of the Sustainable Rural Energy Network (SREN) has developed a concept for the optimization, evaluation and implementation of integrated renewable farms for rural communities under different climatic and environmental conditions (El Bassam 1999).

The concept of an Integrated Energy Community or settlement includes four pathways:

1. Economic and social pathway
2. Energy pathway
3. Food pathway
4. Environmental pathway

Specifications for IEC:

• Decentralized, autonomous and location-based production of energy, food and innovation
• Combined use of different renewable energy sources (biomass, solar, wind, etc.)

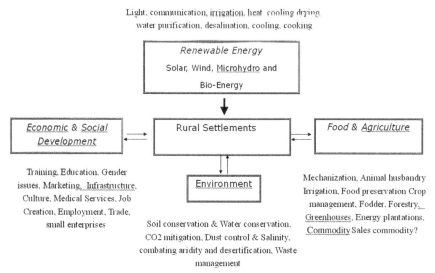

Figure 3.2 *Pathways of the Integrated Energy Community (IEC).* (El Bassam & Maegaard 2004).

- Suitability for remote areas: settlements, villages and islands
- Should include job creation, social, economic, ecological, education and training opportunities for the people in the community

Basic data should be available for the verification of an IREF (Integrated Renewable Energy Farm). Various climatic constraints, water availability, soil conditions, infrastructure, availability of skills and technology, population structure, flora and fauna, common agricultural practices and economic and administrative facilities in the region should be taken into consideration. Renewable energy technologies are available from different natural resources: biomass, geothermal, hydropower, ocean power, solar (photovoltaic and solar thermal), wind and hydrogen. Climatic conditions prevailing in a particular region are the major determinants of agricultural production. In addition, other factors like local and regional needs, availability of resources and other infrastructure facilities also determine the size and the product spectrum of the farmland. The same requisites also apply to an IREF. The climate fundamentally determines the selection of plant species and their cultivation intensity for energy production on the farm. Moreover, climate also influences the production of energy mix (consisting of biomass, wind and solar energies) essentially at a given location and the type of technology that can be installed also depends decisively on the climatic conditions of the locality in question. For example, cultivation of

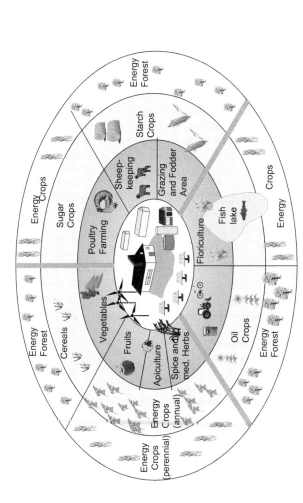

Figure 3.3 *A model for an Integrated Energy Community with farming systems. (El Bassam 2001). (See color plate 7.)*

biomass for power generation is not advisable in arid areas. Instead a larger share can be allocated to solar energy techniques in such areas. Likewise, coastal regions are ideal for wind power installations.

Taking these circumstances into account, a scenario was made for an energy farm of 100 ha (about 247 acres) in the different climatic regions of Northern and Central Europe, Southern Europe, Northern Africa and the Sahara and Equatorial regions (Table 3.1). It was presumed that one unit of this size needs about 200 Megawatt hours (MWh) heat and 100 MWh power per annum for its successful operation. A need for fuel of approximately 8,000 liters per annum has to be calculated. The possible shares of different renewable energies are presented in Table 3.1.

It is evident that, throughout Europe, wind and biomass energies contribute the major share to the energy-mix, while in North Africa and the Sahara the main emphasis obviously lies with solar and wind energies. Equatorial regions offer great possibilities for solar as well as biomass energies and little share is expected from the wind source of energy in these regions. Under these assumptions, in Southern Europe, the Equatorial regions and North and Central Europe, a farm area of 4.8, 10, and 12 ha (12, 24.7, and 29.6 acres), respectively, would be needed for the cultivation of biomass for energy purposes. This would correspond to annual production of 36, 45,

Table 3.1 The possible share of different renewable energy sources in diverse climatic zones produced on an energy farm of 100 Ha (247 acres)

Climate region	Energy source	Power production (% of total need)	Heat production (% of total need)	Biomass need (t/a)	Biomass area (% of total area)
Northern and Central Europe	Solar	7	15	60	12
	Wind	80	—		
	Biomass	60	100		
Southern Europe	Solar	100	100	36	4.8
	Wind	80	—		
	Biomass	40	65		
North Africa Sahara	Solar	100	100	14	1.2
	Wind	100	—		
	Biomass	20	25		
Equatorial region	Solar	100	100	45	10
	Wind	45	—		
	Biomass	70	100		

(Source: El Bassam 1998)

and 60 tons for the respective regions. In North Africa and the Sahara regions, in addition to wind and solar energy, 14 tons of biomass from 1.2% of the total area would be necessary for energy provisions.

3.5.1. Basic Elements of Energy Demand

The IEC concept includes a decentralized living area from which the daily necessities (food and energy) can be produced directly on-site with minimal external energy inputs. The land of an IEC may be divided up into compartments to be used for growing food crops, fruit trees, annual and perennial energy crops and short-rotation forests, along with wind and solar energy units within the farm. An IEC system based largely on renewable energy sources would seek to optimize energetic autonomy and an ecologically semi-closed system, while also providing socio-economic viability and giving due consideration to the newest concepts of landscape and bio-diversity management. Ideally, it will promote the integration of different renewable energies, promote rural development and contribute to the reduction of greenhouse gas emissions.

Energy supply in rural communities has to meet the needs of the people and to ensure economic and social development. In order to generate adequate energy, it is necessary to determine the most appropriate and affordable technologies, equipment and facilities.

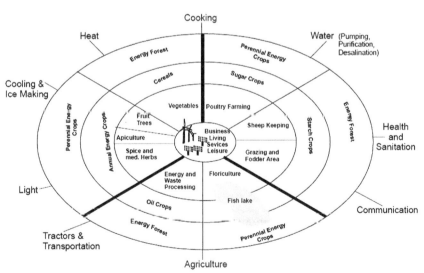

Figure 3.4 *Basic elements and needs of Integrated Renewable Energy Community. (El Bassam 2000).*

3.5.1.1. Heat

Heat can be generated from biomass or solar thermal to create both high temperature steam and low temperature heat for space heating, domestic and industrial hot water, pool heating, desalination, cooking and crop drying.

3.5.1.2. Electric Power

Electric power can be generated by solar PV, solar thermal, biomass, wind, hydro, micro-hydro, geothermal, etc.

3.5.1.3. Water

Water is an essential resource for which there can be no substitute and is needed for drinking and irrigation. Renewable energy can play a major role in supplying water in remote areas. Several systems could be adopted for this purpose:
- Solar distillation
- Renewable energy operated desalination units
- Solar, wind and biomass operated water pumping and distribution systems

3.5.1.4. Lighting

In order to improve living standards and encourage the spread of education in rural areas, a supply of electricity is vital. Several systems could be adopted to generate electricity for this purpose:
- Solar systems (photovoltaic and solar thermal)
- Wind energy systems
- Biomass and biogas systems (engines, fuel cells, Stirling)

3.5.1.5. Cooking

Women in rural communities spend long hours collecting firewood and preparing food. There are other methods that are more efficient, healthy and environmentally benign. Among them are:
- Solar cookers and ovens
- Biogas cooking systems
- Improved biomass stoves using briquettes and pellets
- Plant oil and ethanol cookers

3.5.1.6. Health and Sanitation

To improve serious health problems among villagers, solar energy from photovoltaic, wind and biogas could be used to operate:
- Refrigerators for vaccine and medicine storage
- Sterilizers for clinical items

- Wastewater treatment units
- Ice making

3.5.1.7. Communications

Communication systems are essential for rural development. The availability of these systems has a great impact on people's lives and can advance their development process more rapidly. Electricity can be generated from any renewable energy source to operate the basic communication needs such as radio, television, weather information systems, and mobile telephone.

3.5.1.8. Mobility

Improved transportation in rural areas and villages has a positive effect on the economic situation as well as the social relation between the people of these areas. Several methods could be adapted for this purpose.
- Solar electric vehicles
- Vehicles fueled by biodiesel, ethanol, plant oil, biogas and hydrogen (alternative engines, fuel cells)
- Electric cars
- Traction animals

3.5.1.9. Agriculture

In rural areas agriculture represents a major energy end use. Mechanization using renewable sources of energy can reduce the time spent in labor-intensive processes, freeing time for other income-producing activities. Renewable sources of energy can be applied to:
- Soil preparation and harvesting
- Husking and milling of grain
- Crop drying and preservation
- Textile processing

3.5.1.10. Maintenance Workshops and SM (Small Markets) Industries

Adequate levels of power and heat should be available for processing and manufacturing.

3.6. BASIC AND EXTENDED NEEDS

Power supply systems from renewable sources in off-grid operation should be robust, inexpensive and reliable. Most importantly, they need to have a modular structure so they can be extended later. Photovoltaic (PV) power supply systems in off-grid operation supply power to small consumers (3–30 kW) far from the public utility grid. An essential component of a modular supply system is a battery inverter, such as the Sunny Island, with a nominal power of 3.3 kW each.

The advanced Sunny Island battery inverter is the grid master and the central component of a modular supply system and enables small-scale island utilities in remote areas. An island grid is easy to plan and install and allows a very flexible operation.

The device includes an intelligent control, which is able to supply different consumers and feed power from different generators. Such generators are, for example, PV string inverters for grid supply, small wind energy plants or diesel units, making the battery inverter useful on nysource. The system management handles battery control, enables limited load management and provides communication interfaces for optional system management units. The required operating modes and parallel switching of current converters can be realized. A battery inverter and a lead acid battery can establish a simple single-phase island grid.

3.7. TYPICAL ELECTRICITY DEMANDS

To be able to size the power supply properly, the peak power, the daily energy consumption and the annual energy growth should be estimated. This is necessary in order to size both the generating plant and the conductor used in the distribution system. The sustainability of the system will not be guaranteed if the capacity of the system is too small and leads to consumer dissatisfaction. Alternatively, too much capacity would mean additional investment costs and possibly too high tariffs.

Making load projections that reflect reality is frequently a difficult task to accomplish, especially for prospective consumers who have little experience with electrification. The more reliable approach to assessing demand is to survey households in adjoining, already-electrified areas or in a region with similar economic activities, demographic characteristics, and so forth. This would assess the average initial loads per household in these areas as well as their historical load growth. The already-electrified regions to be surveyed should preferably have a similar type of service as that being proposed in the new community, such as 24-hour power or electricity for 4 hours each evening.

Table 3.2 Daily electricity demand for basic, extended and normal needs

	Basic Needs	Extended Needs	Normal Needs
User Appliances			
Lighting	$3 \times 11W \times 3h$	$4 \times 15W \times 4h$	$4 \times 15W \times 4h$
TV / Radio	$30W \times 4h$	$30W \times 5h$	$60W \times 5h$
Refrigerator		$10W \times 24h$	$30W \times 24h$
Others	100 Wh	300 Wh	1,500 Wh
Central Services for 50 Households			
Water Pumps	3,000 Wh	6,000 Wh	10,000 Wh
Water Treatment	1,500 Wh	3,000 Wh	5,000 Wh
Others	1,500 Wh	1,000 Wh	20,000 Wh
Daily Consumption / Household	440 Wh	1,310 Wh	3,360 Wh
Daily Consumption /	22 kWh	65.5 kWh	168,000 kWh
Average Power	0.92 kW	2.73 kW	7 kW
Peak Power	Ca. 4 kW	11 kW	21 kW

(Source: El Bassam & Maegaard 2004).

Any projections of load and load growth in an area to be electrified using information gathered from already electrified regions should also consider such factors as the difference in the level of disposable income in the two areas, the presence of raw materials or industry, the potential for tourism and access to outside markets for goods which might be grown or produced locally.

For people who are accustomed to the use of electricity, Table 3.2 gives an overview of the daily electricity demand for these three classes: basic, extended and normal needs.

Due to the high cost of electricity generation, it is very important to choose the most efficient appliances. The basis for all assumptions in the table is the use of such appliances. Otherwise the energy consumption will increase significantly. The peak power is typically three to four times higher than the average power.

3.8. SINGLE AND MULTIPLE-PHASE ISLAND GRID

In power generation, PV plants, wind or hydro-electronic power plants can be combined. Normally an additional electric generator such as a diesel generator can make the supply more reliable. The controls allow for a power increase by switching up to three battery inverters in parallel on one phase. In this example, every third household has a refrigerator and every village consists of 50 households.

3.8.1. Version 1: Single-Phase Island Grid

This is a combination of power generation with parallel operation of inverters.

3.8.2. Version 2: Three-Phase Island Grid

If there is the necessity to connect three-phase consumers, the design of the island is flexible and extendable. The smallest three-phase system has a nominal output power of 10 kW and consists of three Sunny Island inverters. Three-phase systems also simplify the connection of larger diesel sets and wind energy systems. These are mostly equipped with three-phase generators.

3.8.3. Version 3: Three-Phase Island Grid and Parallel Operation of the Sunny Island Inverter

Several Sunny Islands can be combined to establish a three-phase system up to 30 kW.

3.8.4. The System Solution for Island Grids

• Simple design of island grids due to connection of all components on the AC side
• Reliable and safe power supply with utility quality in remote areas
• Easy integration of photovoltaic plants and wind energy or diesel generators
• Power supply for single houses or even small villages
• Extendable design (one- or three-phase combinations, parallel operation)
• Optimal battery life

Moving ahead, in order to broaden the scope and seek the practical feasibility of such farms, the dependence of local inhabitants (end users) is to be integrated into this system. Roughly 500 persons (125 households) can be integrated into one farm unit. They have to be provided with food as well as energy. As a consequence, the estimated extra requirement of 1900 MWh of heat and 600 MWh of power have to be supplied from alternative sources. Under the assumption that the share of wind and solar energy in the complete energy provision remains at the same level, the production of 450

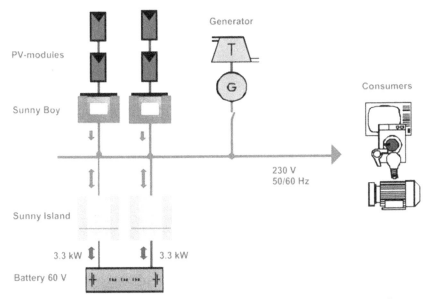

Figure 3.5 *Integration of PV-plants and diesel set and parallel operation of Sunny Island inverters. (El Bassam, N & Maegaard, P 2004).*

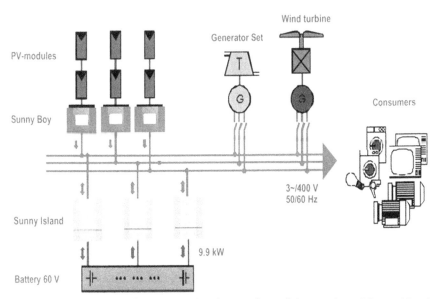

Figure 3.6 *Integration of PV-plants, diesel set and parallel operation of Sunny Island inverters. (El Bassam, N & Maegaard, P 2004).*

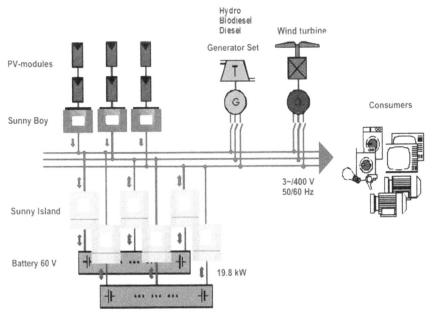

Figure 3.7 *The parallel operation of Sunny Island inverters suitable for high performance. (El Bassam, N & Maegaard, P 2004).*

tons of dry biomass is needed to fulfill the demands of such farm units. For the production of this quantity of biomass, 20% of farm area needs to be dedicated to cultivation. In Southern Europe and the Equator, 15% of the land area should be made available for the provision of additional biomass.

More than 200 plant species have been identified in different regions of the world to serve as sources for biofuels. A summary of energy plant species which can be grown under various climatic conditions is documented in the following sections.

3.8.4.1. Representative Energy Plant Species for Different Climate Regions (Temperate Climate)

(This data is from El Bassam 1996 and 1998b.)

_Cordgrass (*Spartina spp.*)

_Reed Canary Grass (*Phalaris arundinacea.*)

_Fibre sorghum (*Sorghum bicolor*)

_Rosin weed (*Silphium perfoliatum*)

_Giant knotweed (*Polygonum sachalinensis*)

_Safflower (*Carthamus tinctorius*)

_Hemp (*Cannabis sativa)*

_Soybean (*Glycine max*)

_Kenaf (*Hibiscus cannabinus)*

_Sugar beet (*Beta vulgaris*)

_Linseed (*Linum usitatissimum)*

_Sunflower (*Helianthus annuus*)

_Miscanthus (*Miscanthus x giganteus*)

_Switchgrass (*Panicum virgatum*)

_Poplar (*Populus spp.*)

_Topinambur (*Helianthus tuberosus*)

_Rape (*Brassica napus*)

_Willow (*Salix spp.*)

3.8.4.2. Representative Energy Plant Species for Different Climate Regions (Arid and Semi-Arid Climate)

_Argan tree (*Argania spinosa*)

_Olive (*Olea europaea.*)

_Broom (Ginestra) (*Spartium junceum*)

_Poplar (*Populus spp.*)

_Cardoon (*Cynara cardunculus*)

_Rape (*Brassica napus*)

_Date palm (*Phoenix dactylifera*)

_Safflower (*Carthamus tinctorius*)

_Eucalyptus (*Eucalyptus spp.*)

_Salicornia (*Salicornia bigelovii*)

_Giant reed (*Arundo donax*)

_Sesbania (*Sesbania spp.*)

_Groundnut (*Arachis hypogaea*)

_Soybean (*Glycine max*)

_Jojoba (*Simmondsia chinensis*)

_Sweet sorghum (*Sorghum bicolor*)

3.8.4.3. Representative Energy Plant Species for Different Climate Regions (Tropical and Sub-Tropical Climate)

_Aleman Grass (*Echinochloa polystachya*)

_Jatropha (*Jatropha curcas.*)

_Babassu palm (*Orbignya oleifera*)

_Jute (*Crocorus spp.*)

_Bamboo (*Bambusa spp.*)

_Leucaena (*Leucaena leucoceohala*)

_Banana (*Musa x paradisiaca*)

_Neem tree (*Azadirachta indica*)

_Black locust (*Robinia pseudoacacia*)

_Oil palm (*Elaeis guineensis*)

_Brown beetle grass (*Leptochloa fusca*)

_Papaya (*Carica papaya*)

_Cassava (*Manihot esculenta*)

_Rubber tree (*Acacia senegal*)

_Castor oil plant (*Ricinus communis*)

_Sisal (*Agave sisalana*)

_Coconut palm (*Cocos nucifera*)

_Sorghum (*Sorghum bicolor*)

_Eucalyptus (*Eucalyptus spp.*)

_Soybean (*Glycine max*)

_Sugar cane (*Saccharum officinarum*)

3.9. REGIONAL IMPLEMENTATION

The International Research Centre for Renewable Energy in Germany (IFEED) was contracted by the FAO, UN in 2000 to accomplish the planning for the implementation of an IREF in a practical sense at a regional level, taking into consideration the climatic and soil conditions. The planning work was started in Dedelstorf (northern Germany) An area of 280 ha (691.6 acres) was earmarked for this farm, which would satisfy the food and energy demands of the 700 participants in the project. For the settlement purposes, old military buildings were renovated. It was expected to take a period of 3 years to complete the project. The main elements of heat and power generation would be: solar generators and collectors, a wind generator, a biomass combined heat and power generator, a sterling motor and bio-gas plant.

The total energy to be provided was calculated to be as much as 8,000 MWh heat and 2,000 MWh power energy. The cultivation of food and energy crops would be according to ecological guidelines. The energy plant species foreseen were: short-rotation coppice, willow and poplar, miscanthus, polygonum, sweet and fiber sorghum, switch grass and reed canary grass and bamboo. Adequate food and fodder crops as well as animal husbandry would be implemented according to the needs of the people and

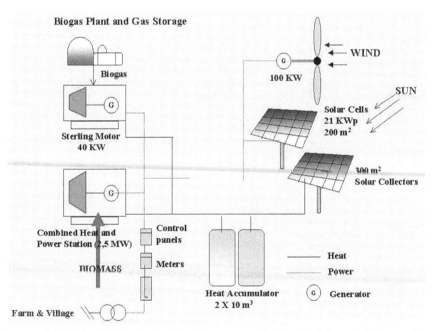

Figure 3.8 *Technologies for heat and power production in Integrated Renewable Energy Community. (El Bassam, N & Maegaard, P 2004).*

specific environmental conditions of the site. A research, training and demonstration center would accompany this project.

In northern Germany a site near Hanover has been already identified for the implementation of an "Integrated Renewable Energy Farm" (90% biomass, 7% wind, 3% solar) and as a research center for renewable energies—solar, wind and energy from biomass as well as their configuration. Special emphasis is dedicated to optimization of energetic autonomy in decentralized living areas and to promote regional resource management. The research center (www.ifeed.org) undertakes the responsibility in the field of research, education, transfer of technology and cooperation with national and international organizations. It also offers trade and industry the opportunity to introduce, demonstrate and commercialize their products. The cooperation with developing countries on the issues of sustainable energy and food production is also one of the prime objectives of the research center.

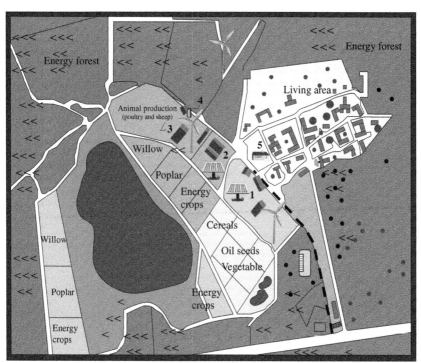

1 Thermal and Power unit (Biomass, Wind, Solar)   2 Pelleting, Oil mill, Ethanol unit
3 Animal husbandry   4 Biogas unit   5 Administration

Figure 3.9 *Implementation of the integrated renewable energy community, Dedelstorf, Germany. (El Bassam 1999).* (See color plate 8.)

REFERENCES

El Bassam, N., 1996. Renewable energy: Potential energy crops for Europe and the Mediterranean region, REU Technical Series 46. Food and Agriculture Organization of the United Nations, Rome (FAO), 200 S.

El Bassam, N., 1998a. Biological Life Support Systems under Controlled Environments. In: Bassam, N El, et al. (Eds.), Sustainable agriculture for food, energy and industry, Volume 2. James & James Science Publishers, London.

El Bassam, N., 1998b. Energy plant species: Their use and impact on environment and development. James & James Science Publishers, London.

El Bassam, N., 1999. Integrated Energy Farm Feasibility Study. SREN-FAO.

El Bassam, N., 2001. Renewable Energy for Rural Communities. Renewable Energy vol. 24, 401–408.

El Bassam, N., Maegaard, P., 2004. Integrated renewable energy for rural communities: Planning guidelines, technologies and applications. Elsevier Science, Amsterdam.

El Bassam, N., 2004. Integrated Renewable Energy Farms for Sustainable Development in Rural Communities. In: Biomass and agriculture. OECD, Paris, pp. 262–276. www.oecd.org/agr/env.

FURTHER READING

Africa Energy Commission (AFREC), 2007.

Business Times of Nigeria, February 10, 2008.

Chiaramonti, D., Grimm, P., Cendagorta, M., El Bassam, N., 1998. Small energy farm scheme implementation for rescuing deserting land in small Mediterranean islands, coastal areas, having water and agricultural land constraints Feasibility study. In: Proceedings of the 10th International Conference. Biomass for energy and industry, Würzburg, pp. 1259–1262.

Davidson, O.R., Sokona, Y., 2001. Energy and sustainable development: Key issues for Africa. In: Wamukonya, N. (Ed.), Proceedings of the African high-level regional meeting on energy and sustainable development. UNEP, Roskilde, pp. 1–20.

CHAPTER FOUR

Planning of Integrated Renewable Communities

4.1. SCENARIO 1

The necessary data such as climatic conditions and soil data, as well as the data concerning the identification of production factors, are normally available for an existing agricultural region. This data should be used as the basis for the planning of Integrated Energy Farms (IEF) and systems.

However, the energy demand of an agricultural area cannot be covered solely by the use of circuit economy—that is, by energetic use of agricultural wastes and residues or through solar and wind energy. In some climatically unfavorable regions we need, in addition to available resources, the production of energy raw materials (biomass) on the farm. However, this may lead to a contentious situation concerning land use between farm products and energy raw materials. The production of the energy crop must not be introduced or expanded at the expense of the other farm products such as foods.

The process of planning an integrated energy system at the community level should include optimizing the farm production in such a way that reserves (land, capital, and manpower) are freed from the field of food production, and can be utilized in an economically feasible way in the process of production of energy and energy raw materials.

A comparison between industrialized and developing countries illustrates that the preferences are very different. In industrialized countries, agriculture is under pressure to reduce food production, while in developing countries food production is of highest importance. Therefore, in the industrialized countries, the replaced production capacity should be used preferably for the production of marketable energy and energy raw materials in order to create new sources of income for the farmers. On the other hand, in the developing countries, the additional production reserves of the farm should only serve to secure energetically autonomous food production.

For planning integrated systems, the following aspects should be first considered:

- Intensification and optimization of existing agricultural production, considering site conditions. This must increase the yields, productivity and income (**optimization model**).
- Choice of adapted production branches with consideration of existing demand and elimination of economically inefficient production branches.
- Introduction of sustainable production models with the aim of minimization of input (capital, surface) in the case of unchangeable net profits.
- Analysis of additional inputs (capital, land, manpower) with regard to their use for additional food production or for production of energy raw material (**preference model**).
- Integration of energy into agricultural production with regard to regional preferences, taking into consideration:
  - criteria of economical profitability
  - technical and economical feasibility
  - availability of technologies and integration possibilities of different technologies
  - surface, capital and manpower requirements
  - socioeconomic and other regional particularities
- Installation of an energy production and user management system.

Figures 4.1 and 4.2 show the different steps of planning and implementation.

4.2. SCENARIO 2

An Integrated Energy Farm could be created on the basis of the energy and food requirements of a specific number of persons. Assuming the farm area is available, the necessary site data such as climate, soil, etc. should be first determined. This can be recorded from national or regional statistics, calculated or measured.

In arid and semi-arid regions irrigation possibilities must be determined. Here, an analysis of the relevant irrigation system and its economic feasibility is very important, taking into consideration the regional sociological aspects.

Figures 4.3 and 4.4 show a graphical presentation of the planning and implementation steps.

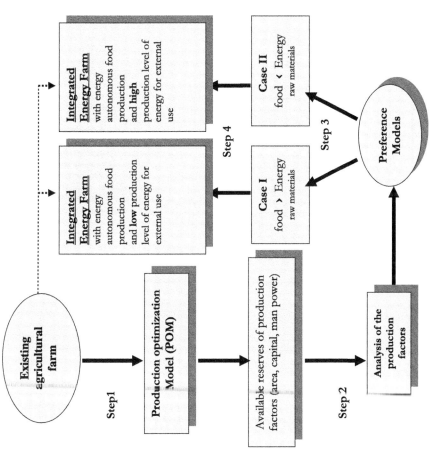

F gure 4.1 *Planning steps Integrated Energy Farm (IEF) (Scenario 1).*

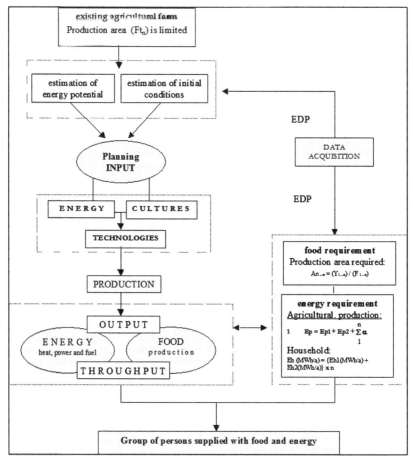

Figure 4.2 *Implementation of Integrated Energy Farm (IEF) (Scenario 1).*

4.3. CASE STUDY I: IMPLEMENTATION OF IEF UNDER CLIMATIC CONDITIONS OF CENTRAL EUROPE

The main planning objective of the implementation of IEF on an existing agricultural farm settlement is that it should be autonomous in energy supply. Specifications are as follows.

4.3.1. Specifications

Climate: Climatic conditions Northern Germany
Farm size: 100 ha (247 acres)
Production: Crops, vegetables, horticulture, energy
                  plantation, animal husbandry, vegetable and fruit production

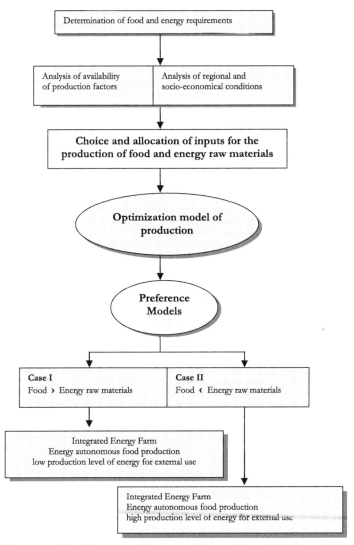

Figure 4.3 *Planning steps of Integrated Energy Farm (IEF) (Scenario 2).*

4.3.2. Distribution of the Farm Area

Crops and root plants:	60 ha (148.2 acres)
Oil plants:	10 ha (24.7 acres)
Vegetables:	5 ha (12.35 acres)
Fruit trees:	5 ha (12.35 acres)
Grassland	17 ha (42 acres)
Building:	3 ha (7.41 acres)
Total:	100 ha (247 acres)

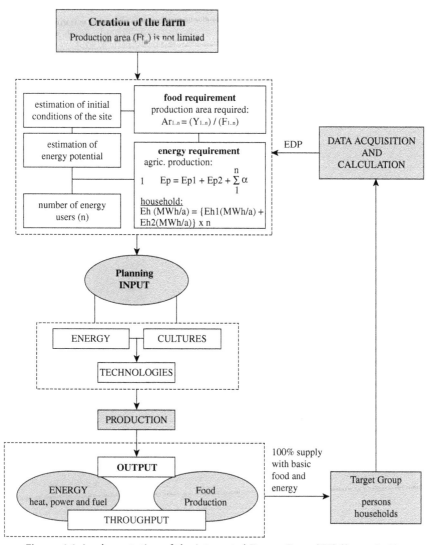

Figure 4.4 *Implementation of the Integrated Energy Farm (IEF) (Scenario 2).*

Figure 4.5a shows a graphical presentation of a general system for power generation of an Integrated Energy Farm.

The infrastructure of the farm might consist of a residential building with administration tract, office rooms (area: 400 m^2), storage rooms, workshops, greenhouses and stable facilities.

Figure 4.5b shows the basic diagram of the system control unit of the power generation system on an Integrated Energy Farm.

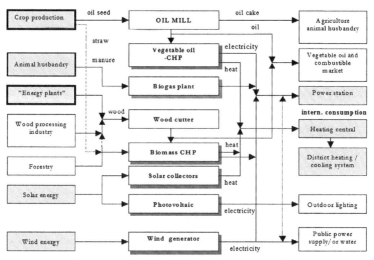

Figure 4.5a *General system complex of power generation of an Integrated Energy Generation.*

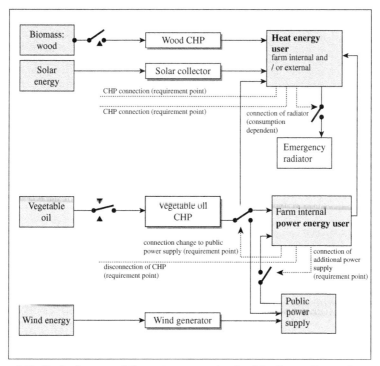

Figure 4.5b *Basic diagram of the system control unit of the Power Generation system on an Integrated Energy Farm. (Source: Wolf, M 1998, modified by the authors).*

4.3.3. Farm Production

The activities of the created Integrated Energy Farm consist of:

Crop production: 70 ha (172.9 acres)
Energy plantation: 15 ha (37.1 acres)
Grassland /fodder: 2 ha (4.94 acres)
Horticulture
- Vegetables: 5 ha (12.35 acres)
- Fruit trees: 5 ha (12.35 acres)

Animal husbandry
- Sheep breeding (No.) 100
- Chicken (No.) 500

Table 4.1 presents the production data of the different agricultural activities.

Table 4.1 Agricultural production per year (estimated)

Production branch	Products	Area	Quantity
Crop production	Cereals (wheat and barley)	40.0 ha	100 t
	Sunflower	10.0 ha	15 t
	Potatoes	20.0 ha	3000 t
Vegetables	Different products	5.0 ha	100 t
Fruit trees	Apples, pears, etc.	5.0 ha	200 t
Animal	Hens (eggs)	—	75.000 units
Husbandry	Sheep breeding*	2.0 ha	2.0 t LW

* 1.0 Lamb per year, LW = Life Weight, dt = 0.1 ton

4.3.4. Energy Requirement

4.3.4.1. Administration and Household

The area for the living and administration buildings is projected to be approximately 400 m^2.

The determination of the **heat energy requirement** is based on the following criteria:

- The farm buildings should be well insulated so that relatively low energy for heating will be required, corresponding to 60–70 watts per m^2 reference area.
- The total hours needed for heating will be approximately 1800 h/y (10 hours/d × 6 months /y × 30 days).

$$Heat\ requirement\ (kWh/y) = used\ area\ (m^2) \times 70W/m^2 \times 1,800\ h$$

$$= 400 \times 70 \times 1800 \times 10^6 = 50.4 MWh/y$$

Table 4.2 Power requirement in MWh/Year

Fields	Connection value (kW)	Full load hours (h)	Consumption (MWh/y)
Basic load	2.5	8,760	21.9
Household	2.5	1,100	2.75
Cooking	8.2	730	5.99
Administration	1.5	1,300	1.95
Workshop	12.1	450	5.45
Sum	**26.8**		**38.04**

The hot water requirement is calculated to be 40 l/d/person. The annual demand for energy for hot water from 20 to 60°C is calculated as follows:

$$40 \times (60 - 10) \times 4.19 \times 365 \times 0.000278 = 850.3 kWh/y/person$$

For the entire building, a heating load of approximately **28.0 kW** should be available, so that the entire annual energy demand will be 50.4 **MWh/year**.

Table 4.2 illustrates the electricity needed for various activities in administration and household (six persons):

A basic load of 2.5 kW for the entire farm is projected for permanent (24 h/day) power generation for circulating pumps, ventilation, engines, refrigerator, emergency lighting, etc.

The total electricity consumption amounts to 38.1 **MWh/year**; this corresponds to a consumption of **104.2 KWh** per day.

4.3.4.2. Agricultural Activities

Energy requirements, subdivided into power, heat and fuel for the various farming activities, are summarized in Table 4.3.

For all farming activities 824 tractor hours per year are needed with a total fuel consumption of 6476 liters (about 1711 US gallons). The requirements for electricity and heat energy are calculated to be 39.5 MWh/y and 75.0 MWh/y, respectively.

Summarizing the energy demand of the farm production and that of the household and administration space, the following energy is required for the entire farm:

Heating energy: 125.4 MWh/y

Electricity: 77.6 MWh/y

Fuel: 6476.2 liters

Table 4.3 Electricity, heat energy and fuel requirement on the farm

Branches	Activities	Area in ha	Th */area	Electricity (MWh)	Heat (MWh)	Fuel (liters)
Cereals	soil	40.0	144.0	—	—	1166.4
	preparation		0	—	—	583.2
	care		72.0	—	—	81.0
	harvest		10.0	—	—	518.4
	straw		64.0	—	—	
	transport/recovery	19.0		—	—	76.5
Potatoes			95.0	—	—	769.5
Oil plants	soil		95.0	5.5	—	738.7
	preparation		91.2			
	care	10.0		—	—	542.7
	harvest/storage		67.0	—	—	129.6
			16.0	9.8	—	162.0
			20.0	7.5	—	—
Vegetables	soil		—	—	—	—
	preparation	5.0				52.7
	care		6.5	—	—	81.1
	harvest/storage		10.0	7.4	—	—
	processing	5.0	—	—	—	60.8
Fruit trees	care		7.5	—	—	48.6
	harvest		6.0	3.5	—	—
	processing		—			
	care					
	harvest/transport					
	port					
	storage					
Sum 1	-	**79.0**	**704.2**	**33.7**	**—**	**5704.2**
Animal husbandry	Feeding other	5.0	120.0	5.8	75.0	972.0
Sum 2		**5.0**	**120.0**	**5.8**	**—**	**972.0**
Total		84.0	824.2	39.5	75.0	6476.2

* Th = tractor hour

4.3.4.3. Site Energy Production

The total energy demand calculated for heat was 125.4 MWh/y and for electricity 77.6 MWh/y. The required energy should be produced fully on the farm. Surplus electric energy can be sold to the public energy network.

The energy farm exploits exclusively renewable energy resources. The concept includes a combined use of solar and wind energy as well as the energy from biomass. The energy supply system of the farm consists of the following technologies:

- **Photovoltaic plant** of approx. 5.3 kW (50 modules with an output capacity of 105 Wp/module on a surface of 50 m^2)
- **Thermal solar collectors** with 100 m^2 collecting surface
- **Windmill** of 300 kW
- **Wood/Biomass-CHP**, 100 kW thermal capacity
- **Stirling engine** with 10 kW electrical and 20 kW of thermal capacity that is operated with biogas
- **Biogas plant** having a production capacity of 750–1000 m^3 biogas/year

Table 4.4 shows the contribution of various technologies to produce energy from renewable sources.

4.3.4.4. Origin of Biomass

The biomass for energy generation is provided on the farm: 15 ha (37 acres) will be cultivated with fast-growing tree species, such as eucalyptus and poplars, as well as various energy plants such as giant reed, miscanthus, different tall grasses, etc.

Consequently, the farm utilizes about 90 tons of biomass annually as energy raw material producing approximately 400 MWh/y.

The biogas plant is operated with animal manure, plant wastes and farm residues (estimated quantity about 40 tons/y) producing 700–750 m^3 biogas.

A Stirling engine operated with biogas produces 4.4 MWh/y heat and 1.6 MWh/y of electricity.

Table 4.4 Estimated annual energy production on the farm

	Heat energy	Electricity
Photovoltaic plant	—	5.0 MWh [1]
Thermal solar collector	40.0 MWh	—
Windmill	—	120.0 MWh [2]
CHP (cogeneration)	260.0 MWh	125.0 MWh
Stirling engine (with biogas)	4.4 MWh	1.6 MWh
Total	304.4 MWh	251.6 MWh

[1]Calculated on the basis of an average global sun radiation of 2.7 KWh/m^2/d and its energetic use of 10–12%.
[2]Wind speed on the site = 4.5–5.0 m/sec^3) annual operation hours = 2100 h.

The total quantity of energy generation from the produced biomass on the farm amounts to 304.4 MWh/y heat energy.

4.3.4.5. Contribution of Different Renewable Energy Sources

Table 4.5 shows the contribution of different energy sources by percentage to cover the total energy requirement of the farm (heat: 377.4 MWh/y, power: 107.6 MWh/y, fuel: 7093.4 liters).

Table 4.5 Energy demand and energy generation

	Solid biomass	Oil plants	Biogas	Solar-energy	Wind-energy	Total	% of total demand
Heat	400.0 MWh	—	4.4 MWh	40.0 MWh	—	444.4	117.8
Electricity	—	—	1.6 MWh	7.5 MWh	200.0 MWh	209.1	194.4
Fuel	—	2,500 l	—	—	—	2,500 l	35.2

4.3.4.6. Investment Requirement

The investment requirement has been estimated on the basis of information collected from different professional organizations, energy agencies and energy producers. The total sum of the required investment, including all capital and additional costs such as financing and services costs, amounts to about 918,000 EUR (about $1,162,188.00 US).

In this case, the estimated total investment requirement consists of the following costs of equipment, installation and service:

BOX 4.1 Estimated Total Investment Costs

Windmill (300kW):	300,000
Photovoltaic solar cells for power generation (50 units with the capacity of 105Wp/units) +installation:	38,000
Solar collectors for heat energy (100 m²):	15,000
Biomass - CHP:	125,000
Biogas plant with reservoir and generator:	75,000
Oil mill + tank + Installation:	110,000
Wood cutter + Container:	35,000
Plantation of energy plants:	125,000
Financing costs, planning and service etc.:	95,000
Total EUR:	**918,000**

4.4. CASE STUDY II: ARID AND SEMI-ARID REGIONS

The main objectives of the planning are the implementation of a farm settlement that is autonomous in energy supply including desalination and cooling facilities from renewable energy sources.

4.4.1. Specifications

Climate: Arid and Semi-Arid Regions and Islands

Farm size: 100 ha

Production: crops, vegetables, horticulture, energy plantation, animal husbandry, pisciculture and apiculture

---

BOX 4.2 Distribution of Farm Area

Distribution of the farm area

Crops and root plants:	60 ha
Oil plants:	10 ha
Vegetables:	5 ha
Fruit trees:	5 ha
Grassland:	17 ha
Building:	3 ha
Total:	**100 ha**

The infrastructure of the farm might consist of a residential building with administration space, office rooms (using area: 400 m^2), storage rooms, workshops, greenhouses and stable facilities.

4.4.2. Farm Production

Table 4.6 presents the production data for different branches.

BOX 4.3 The Activities of the Created Integrated Energy Farm Consist of:

Crop production:	70 ha
Energy plantation:	15 ha
Grassland /fodder.	2 ha
Horticulture	
• Vegetables:	5 ha
• Fruit trees:	5 ha
Animal husbandry	
• Sheep breeding (No.)	100
• Chicken (No.)	500

Table 4.6 Agricultural production per year (estimated)

Production branch	Products	Area	Quantity
Crop production	Cereals (Wheat and barley)	40.0 ha	100 t
	Sunflower	10.0 ha	15 t
	Potatoes	20.0 ha	3000 t
Vegetables	Different products	5.0 ha	100 t
Fruit trees	Apples, Pears, etc.	5.0 ha	200 t
Animal husbandry	Hens (eggs)	—	75,000 units
	Sheep breeding*	2.0 ha	2.0 t LW

* 1.0 Lamb per year, LW = Life Weight, dt = 0.1 ton

4.4.3. Energy Requirement
4.4.3.1. Administration and Household
The area for the living and administration buildings is projected to be approximately 400 m^2.

The determination of **cooling load requirement** is based on the following criteria:

- The farm buildings should be well insulated so that relatively low energy for cooling will be required which corresponds to 1 ton per 20 m^2 (3.5 kW/ton).
- The total hours needed for cooling will be approximately 4320 h/a (18 hours/d × 8 months/y × 30 days).

BOX 4.4 Formula for Cooling

Cooling load requirement (MWh/y) = area (m^2) × 3.5 kW/ 20 × 4320 h

$-$ 400 × 3.5/20 × 4320 = 302.4 MWh/y of heat energy to be used in absorption chillers to produce chilled water for cooling purposes

Hot water requirement is calculated to be 40 l/d/person. The annual demand for energy for hot water from 20 to 60 $_o$C is calculated as follows:

BOX 4.5 Formula for Hot Water

$$40 \times (60 - 20) \times 4.19 \times 365 \times 0.000278 = 783.0 \, kWh/y/person$$

For all the buildings, a cooling load of approximately **75.0 kW** should be available, so that the entire annual energy demand will be **302.4 MWh/year**.

Table 4.7 Power requirement in MWh/year

Fields	Connection value (KW)	Full load hours (h)	Consumption (MWh/y)
Basic load	2.5	8,760	21.90
Household	2.5	1,100	2.75
Cooking	8.2	730	5.99
Administration	1.5	1,300	1.95
Workshop	12.1	450	5.45
Sum	**26.8**	**12,340**	**38.04**

Table 4.7 indicates the electricity needed for various activities in administration and household (6 persons).

The basic load of 2.5 kW for the entire farm is projected for permanent (24 h /day) power generation for circulating pumps, ventilation, engines, emergency lighting, refrigerators, etc.

The total electricity consumption amounts to **38.1 MWh/year**, corresponding to a consumption of **104.2 KWh** per day.

4.4.3.2. Agricultural Activities

Energy requirement, subdivided into power, heat and fuel for the various farming activities, is summarized in Table 4.8.

For all farming activities 24.2 tractor hours per year are needed with a total fuel consumption of 6476.2 liters. The requirement of electricity and heat energy is calculated to be 39.5 MWh/a and 75.0 MWh/y, respectively.

Summarizing the energy demand of the farm production and that of the household and administration space, we will have the following energy requirement situation for the entire farm:

heating energy: 377.4 MWh/y

electricity: 107.6 MWh/y

fuel: 7093.4 liters

4.4.3.3. Energy Production on the Farm

The total energy demand was calculated for heat at 377.4 MWh/y and for electricity 107.6 MWh/y. The required energy should be produced fully on the farm. The surplus electric energy will be used to power the desalination system.

The energy farm should exploit exclusively renewable energy resources. The concept includes a combined use of solar and wind energy as well as the energy from biomass.

Table 4.8 Electricity, heat energy and fuel requirement on the farm

Branches	Activities	Area in ha	Th */ area	Electricity (MWh)	Heat (MWh)	Fuel (liter)
Cereals	Soil	40.0	144.0	—	—	1166.4
	Preparation			—	—	583.2
	Care		72.0	—	—	81.0
	Harvesting		10.0		—	518.4
	Straw		64.0	—	—	
Potatoes		19.0			—	769.5
	Transport/ recovery		95.0	—	—	769.5
	Soil		95.0	5.5	—	738.7
Oil plants	Preparation	10.0	91.2	—	—	542.7
	Care		67.0	—	—	129.6
	Harvest/storage		16.0	9.8	—	162.0
			20.0	7.5	—	—
Vegetables	Soil	5.0	—	—	—	52.7
	Preparation		6.5	—	—	81.1
	Care		10.0	7.4	—	—
Fruit trees	Harvest/storage	5.0	—	—	—	60.8
			7.5	—	—	48.6
	Processing		6.0	3.5	—	—
Energy	Care	15.0	—			
Plantation	Harvest	2.0				
Grassland	Processing					
	Care					
	Harvest/transport					
	Storage					
Sum 1	—	**79.0**	**704.2**	**33.7**	—	**5704.2**
Animal	Feeding	5.0	120.0	—	—	972.0
Husbandry	Other	—		5.8	75.0	—
Sum 2	—	**5.0**	**120.0**	**5.8**	—	**972.0**
Total		84.0	824.2	39.5	75.0	6476.2

* Th = tractor hour

The energy supply system of the farm consists of the following technologies:

- **Photovoltaic plant** of approx. 5.3 kW (50 modules with an output capacity of 105 Wp/module on a surface of 50 m^2)
- **Thermal solar collectors** with 100 m^2 collecting surface
- **Windmill** of 300 kW

- **Wood/Biomass-CHP** of 100 kW thermal capacity
- **Stirling engine** with 10 kW electrical and 20 kW of thermal capacity that is operated with biogas
- **Biogas plant** having a production capacity of 750–1000 m³ biogas/year

Table 4.9 shows the contribution of various technologies to produce energy from renewable sources.

Table 4.9 Estimated annual energy production on the farm

	Heat energy	Electricity
Photovoltaic plant	—	7.5 MWh 1)
Thermal solar collector	40.0 MWh	—
Windmill	—	200.0 MWh 2)
CHP (cogeneration)	360.0 MWh	3)
Stirling engine (with biogas)	4.4 MWh	1.6 MWh
Total	**404.4 MWh**	**208.1 MWh**

[1]Calculated on the basis of an average global sun radiation of 5.0 kWh/m²/d and its efficiency of 10–12%.
[2]Wind speed on the site = 3.5–4.0 m/sec.
[3]Annual operation hours for CHP = 3100 h.

4.4.3.4. Origin of Biomass

The biomass for energy generation is provided on the farm; 15 ha will be cultivated with fast-growing tree species such as eucalyptus and poplars, as well as various energy plants such as giant reed, miscanthus, different tall grasses, etc.

Consequently, the farm disposes of about 90 tons of biomass annually as energy raw material producing approximately 400 MWh/y.

The biogas plant operates with stable manure, plant wastes and farm residues (estimated quantity about 40 tons/a) producing 700–750 m³ biogas.

A Stirling engine operated with biogas produces 4.4 MWh/y heat and 1.6 MWh/y of electricity.

The total quantity of energy generation from the produced biomass on the farm amounts to 404,4 MWh/y heat energy.

4.4.3.5. Contribution of Different Renewable Energy Sources

Table 4.10 shows the percentage contribution of different energy sources to cover the total energy requirement of the farm (heat: 377.4 MWh/y, Power: 107.6 MWh/y, fuel: 7093.4 liter).

Table 4.10 Energy demand and energy generation

	Solid biomass	Oil plants	Biogas	Solar-energy	Wind-energy	Total	% of total demand
Heat	400.0 MWh	—	4.4 MWh	40.0 MWh	—	444.4	117.8
Electricity	—	—	1.6 MWh	7.5 MWh	200.0 MWh	209.1	194.4
Fuel	—	2,500 l	—	—	—	2,500 l	35.2

4.4.3.6. Investment Requirement

The investment requirement has been estimated on the basis of information collected from different professional organizations, institutions of energy supply and producers. The total sum of the required investment, including all capital and additional costs such as financing and services, amounts to about **1,118,000 EUR (about $1,415,388 US)**.

BOX 4.6 In This Case, the Estimated Total Investment Requirement Consists of the Following Costs of Plants, Installation and Service:

1. Windmill (300kW):	300,000
2. Photovoltaic solar cells for power generation (50 units with the capacity of 105Wp/units) +Installation:	38,000
3. Solar collectors for heat energy (100 m²):	15,000
4. Biomass - CHP:	125,000
5. Biogas plant with gas storage and generator:	75,000
6. Oil press + tank + Installation:	110,000
7. Wood cutter + container:	35,000
8. Plantation of energy plants:	125,000
9. Financing costs, planning and service, etc. :	9,000
10. Reverse osmosis desalination unit (capacity 10 m³/h)	100,000
11. Absorption chillers and other air conditioning equipment	50,000
12. Irrigation equipment	50,000
Total : EUR	**1,118,000**

REFERENCE

El Bassam, N., Maegaard, P., 2004. Integrated renewable energy for rural communities: Planning guidelines, technologies and applications. Elsevier Science, Amsterdam. 50–70.

CHAPTER FIVE

Determination of Community Energy and Food Requirements

5.1. MODELING APPROACHES

The modeling procedure should include the identification and determination of the following parameters:

1. Site conditions:
   - climate: temperature, amount and distribution of precipitation, sunshine duration, wind velocity (annual mean).
   - soil conditions, irrigation possibilities, etc.
   - factors of production: capital, machines, building, agricultural area.
2. Energy requirement per year for food production for households
3. Basic food requirement per person per year
4. Number of energy consumers (persons and households)
5. Site energy potential: solar energy, wind energy and biomass
6. Preparation of a master production schedule for food and energy production
7. Selection and installation of suitable technical tools using the renewable energy resources of the site
8. Energy production and use management
9. Environmental impact
10. Social and economic impact

For elaboration, development and establishment of the Integrated Energy Farm (IEF), two scenarios have been considered:

5.1.1. Scenario 1 (Figure 5.1)

Initial conditions: The farm size (Ft_n) is known; that is, an already existing farm, or the agricultural area is limited.

Objectives: The available area (Ft_n) should be managed to achieve a high degree of self-sufficiency for a maximum number of people with basic food (N_x) and energy.

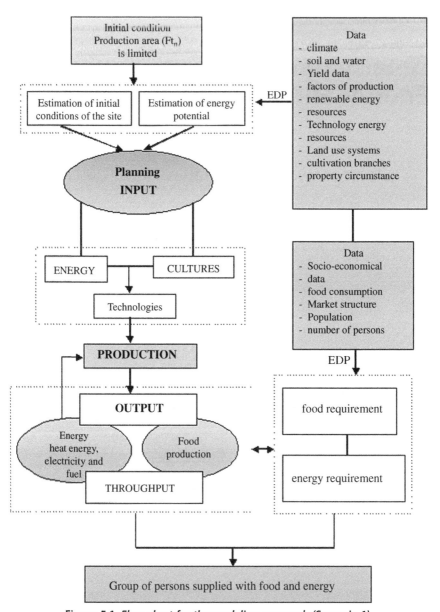

Figure 5.1 *Flow chart for the modeling approach (Scenario 1).*

5.1.2. Scenario 2 (Figure 5.2)

Initial conditions: The farm size (Ft$_x$) is variable; that is, the size could be adapted according to needs.

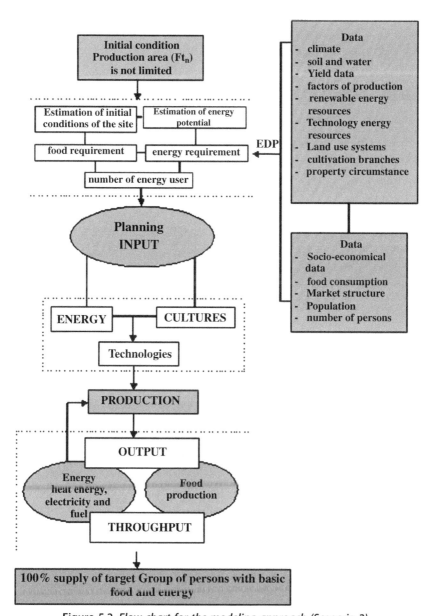

Figure 5.2 *Flow chart for the modeling approach (Scenario 2).*

Objectives: High degree of self-sufficiency for a determined number of people with basic food and energy should be achieved.

5.2. DATA ACQUISITION

The data shown in Table 5.1 should be identified in details for the planning, modeling and implementation of an Integrated Energy Farm.

5.3. DETERMINATION OF ENERGY AND FOOD REQUIREMENTS

5.3.1. Agricultural Activities

Agriculture is itself an energy conversion process, namely the conversion of solar energy through photosynthesis to food energy for humans and feed for animals. Primitive agriculture involved little more than scattering seeds on the land and accepting the scanty yields that resulted. Modern agriculture requires an energy input at all stages of agricultural production such as direct use of energy in farm machinery, water management, irrigation, cultivation

Table 5.1 Data acquisition (Overview)

	To be measured	To be calculated	To be recorded
External Data:			
Climate data:			
Precipitation and distribution	X		X
Temperature	X		X
Annual temperature variations			X
Socio–economic data:			
Size of the population			X
Property structures			X
Age structure of the population			X
Education			X
Economy data, number of trade companies, industry			X
Employment situation			X
Land use: agriculture and forestry			X
Land division: crop production, animal breeding, fruit cultivation, vegetable, pisciculture, forestry area, etc.			X
Number of households			X

and harvesting. Post-harvest energy use includes energy for food processing, storage, and transport to markets. In addition, there are many indirect or sequestered energy inputs used in agriculture in the form of mineral fertilizers and chemical pesticides, insecticides and herbicides.

While industrialized countries have benefited from these advances in energy availability for agriculture, developing countries have not been so fortunate. "Energizing" the food production chain has been an essential feature of agricultural development throughout recent history and is a prime factor in helping to achieve food security. Developing countries have lagged behind industrialized countries in modernizing their energy inputs to agriculture.

Agriculture accounts for only a small proportion of total final external commercial energy demand in both industrialized and developing countries. In the OECD countries, for example, around 3–5% of commercial energy consumption is used directly in the agricultural sector. In developing countries, estimates are more difficult to find, but the equivalent figure is likely to be similar—in the range of 4–8% of total final commercial energy use.

The data for non-renewable energy use in agriculture also excludes the energy required for food processing and transport by agro-industries. Estimates of these activities range up to twice the energy reported solely in agriculture. Definitive data does not exist for many of these stages, and this is particularly problematic in analyzing developing country energy statistics. In addition, the data conceals how effective these energy inputs are in improving agricultural productivity. It is the relationships between the amounts and quality of the direct energy inputs to agriculture and the resulting productive output that are of most interest.

Looking more closely at energy use in specific crops, comparisons of commercial energy use in agriculture for cereal production in different regions of the world are listed in Table 5.2. The relationship between commercial energy input and cereal output per hectare for the main world regions is also shown in Figure 5.3. These data, while relatively old, indicate that developing countries use less than half the commercial energy input (whether in terms of energy per hectare of arable land or energy per ton of cereal) compared with industrialized countries. However, this is not to say that developing countries are necessarily more efficient in their use of energy for agricultural production.

A comparison between the commercial energy required for rice and maize production by modern methods in the United States and transitional

Table 5.2 Commercial energy use and cereal output (1982)

Region	Energy per hectare of arable land (kgoe/ha)	Energy per ton of cereal (kgoe/t)	Energy per agricultural worker (kgoe/person)
Africa	18	20	26
Latin America	64	32	286
Far East	77	43	72
Near East	120	80	285
All developing countries average	96	48	99
All industrialized countries average	312	116	3294
World average	**195**	**85**	**344**

(Source: Stout, 1990).

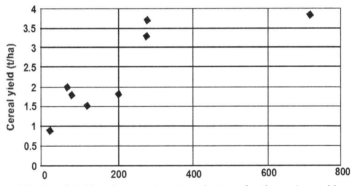

Figure 5.3 *Cereal yield and energy input per hectare for the main world regions.*

and traditional methods used in the Philippines and in Mexico is shown in Table 5.3. The data shows that the modern methods give greater productive yields and are much more energy-intensive than transitional and traditional methods. These methods include the use of fertilizer and other chemical inputs, more extensive irrigation and mechanized equipment.

The energy data for production **(Ep)** depends first on the size of the farm **(Sq)**, the mechanization degree and the production activities. There is the requirement of fuel **(Ep1)** and of electricity **(Ep2)**.

Fuel is required for activities of soil preparation and cultivation, as well as for harvesting and transportation.

BOX 5.1 Fuel Requirement (Ep1) Is Calculated as Follows:

$Ep1(1/y) =$ working duration/ha $\times$ area(Sq1) $\times$ fuel consumption/h/Machine

$$Ep1(liter/year) = \sum Sh/ha \times Sq1 \times al \qquad (1)$$

Sh = working duration per machine and per ha
Sq1 = size of the field
Al = fuel consumption per machine and per hour
1...n = different cultivation branches
ha = hectare

Table 5.3 Rice and maize Production by modern, transitional and traditional methods

	Rice production			Maize production	
	Modern (United States)	Transitional (Philippines)	Traditional (Philippines)	Modern (United States)	Traditional (Mexico)
Energy input (MJ/ha)	64,885	6,386	170	30,034	170
Productive yield (kg/ha)	5,800	2,700	1,250	5,083	950

(Source: Stout 1990).

The electricity requirement depends on the mechanization degree. Electricity is required especially in animal production but also for the storage, cooling and drying of crops. The energy requirement for

---

BOX 5.2 Energy Requirement for Production (Ep2):

$$Ep2(MWh/y) = \sum h/y \times al1 \qquad (2)$$

h/a = working duration per machine and per year
al_1 = power requirement per machine and per hour
1...n = different cultivation branches

> **BOX 5.3 Considering Formula (1) and (2) the Total Energy Requirement for Agricultural Production Is to Be Calculated as Follows:**
>
> $$Ep = Ep1 + Ep2 + \sum \alpha$$
>
> α = losses in kWh/a/machine

production **(Ep2)** is the sum of all usage factors of all electric instruments and machines to be used on the farm.

5.3.2. Households

The energy requirement for the households **(Eh)** is divided into heat energy (for heating, cooling and hot water preparation) and the power requirement for light, electrical appliances and for cooking. The requirement data are different from region to region depending on climatic conditions and should be calculated considering the specific site conditions.

5.3.2.1. Heat Energy

The requirement data for heating and cooling load **(Eh1/household)** depends in addition on the site and environmental conditions and also on the construction type of the buildings (such as full insulation, half insulation).

Hot water requirements are indicated considering different use possibilities in liter/day/person.

> **BOX 5.4 Heating and Cooling Loads for the Buildings Are Calculated as Follows:**
>
> $$Eh1(MWh/y) = \gamma(W/m^2) \times h/y \times F(m)^2 \qquad (5.3)$$
>
> $\gamma(W/m^2)$ = energy requirement per reference area
> h/y = year requirement of energy in full load hour
> F (m^2) = Energy reference area

BOX 5.5 The Total Heat Energy Requirement for Hot Water with a Temperature of 60°C Is Determined as Follows:

$$Q(kJ/y) = mX(t_2 - t_1)X\ 4.19\ kJ/KX\ 365$$

$$Q(kJ/y)X\ 0.000278 = Q(kWh/y)$$

Q = heat energy requirement
m = hot water consumption per day (l)
t1 = cold water temperature (10 − 20°C)
t2 = desired temperature (60°C

5.3.2.2. Electricity

BOX 5.6 The Total Requirement (Eh2/Household) Results from the Usage Data of All Electrical Devices Existing in a Household for Different Uses: Light, Communication, Cooking and Cooling.

$$Eh2(MWh/y) = \sum \beta(kW)x\ h + \alpha(kWh) \qquad (5.4)$$

$\beta(kW)$ = connected load of the device
h = full load hours in the year
α = losses in **kWh**

BOX 5.7 Considering Equation (5.3) and (5.4) the Total Energy Requirement of the Household Is Calculated as Follows:

$$Eh(MWh/y) = \{Eh1(MWh/y) + Eh2(MWh/y)\} \times n$$

$$n = number\ of\ households$$

5.3.3. Food Requirement

On a broad regional basis, there appears to be a correlation between high per capita modern energy consumption and food production. Figure 5.4 shows

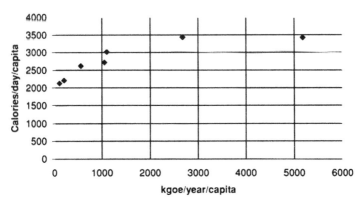

Figure 5.4 *Modern energy consumption and food intake.*

Table 5.4 Basic foods and their sources

Food components	Sources	Indicators
Carbohydrates	Cereals	kg/person/year
	Other crops	kg/person/year
Protein	Animal	Number of animals/farm
	Plant	kg/person/year
Vitamin	Fruits	kg/person/year
	Vegetables	kg/person/year
Fats	Oil plants	kg/person/year
	Animal	Number of animals/farm

data for daily food intake per capita and the annual commercial energy consumption per capita in seven world regions (FAO, 1995)[*].

While broad data on a regional basis conceal many differences between countries, crop types and urban and rural areas, the correlation is strong in developing countries, where higher inputs of modern energy can be assumed to have a positive impact on agricultural output and food production level. The correlation is less strong in industrialized regions where food production is near or above required levels and changes in production levels may reflect changes in diet and food fashion rather than any advantages gained from an increased supply of modern energy.

Basic food requirement (person/year) is divided into requirement data for carbohydrates, proteins, vitamins and fats. They are different from

[*] The regions are: Sub-Saharan Africa Asia, East Asia/Pacific, Latin America/Caribbean, Middle East/North Africa, Europe, OECD.

country to country. For the realization of an Integrated Energy Farm and for the preparation of a land use plan, the regional and national consumer data should be used. In Table 5.4, the basic foods and their resources from agricultural production are performed.

The data for food consumption are normally known and must be used while planning. Which kind of food will finally be produced on the farm depends on the climatic and soil conditions of the site.

BOX 5.8 For the Determination of the Production Area for Each Agricultural Branch the Following Calculation Formula Can Be Used:

$$Ar_{1...n} = (Y_{1...n})/(F_{1...n})$$

Ar = space requirement in ha
Y = yield/area
F = requirement/head/year
1...n = different basic food

5.4. ENERGY POTENTIAL ANALYSIS

5.4.1. Solar Energy

The total solar radiation that strikes the earth's surface amounts to 1018 kWh/a, which is many times greater than the present global energy demand.

In the case of vertical incidence (solar altitude: 90°), the radiation intensity of global radiation can reach a level of 1100 W/m^2.

The daily sum of global radiation (horizontal surface) on a sunny day in the vicinity of the equator is estimated to be 6–8 kWh/m^2 / d.

It is possible to estimate the proportion of global radiation represented by diffuse radiation (important for the utilization of solar collectors). It is a function of the solar altitude and the degree of cloudiness and ranges from 10–85%.

The available data regarding regional solar radiation come from measurements on horizontal surfaces. But to increase their efficiency, the solar collectors are usually mounted at a tilt angle. The solar radiation is divided into two components: the direct radiation and the diffuse radiation. The conversion of the direct radiation is relatively simple. However, specific assumptions must be made for conversion of the diffuse component, because this radiation component is extremely dependent on site conditions and on technical facilities. Generally three estimating procedures are possible:

1. The first assumes that the sky radiation becomes the predominant part of direct solar proximity. This can be possible only on clear days.

BOX 5.9 The Total Radiation Falling on a Tilt Surface $G_{G,g}$ Is Then:

$$G_{G,g} = RG_{G,h}(W/m^2)$$

R = ratio of the direct radiation on a tilt surface ($G_{o, g} = G_o \cos \varphi$) to the radiation value on a horizontal surface

---

2. The second assumes that the sky radiation is distributed uniformly via the entire sky vault. This is an approximation for cloudy days.

BOX 5.10 The Entire Radiation Falling on a Tilt Surface Results in:

$$G_{G,g} = RG_{D,h} + RG_{H,h}(W/m^2)$$

$G_{D,h}$ = direct radiation on horizontal surface
$G_{H,h}$ = sky radiation on horizontal surface

---

3. The third procedure represents a middle way between the two extremes. One assumes here that, because of the inclination of the absorber, it sees only a part of the sky vault (indeed ½ (1+cos n), but it receives an additional diffuse part of the radiation in the form of ground reflection from the collector environment.

BOX 5.11 Total Radiation Falling on a Tilt Surface Consists of Three Parts: Direct Component, Diffuse Component and Reflected Part:

$$G_{G,g} = RG_{D,h} + [1/2(1 + \cos n)]G_{H,h} + \sigma_B G_{G,h}(W/m^2)$$

n = inclination angle of the receiving face (degree)
σ_B = reflection coefficient of the surrounding ground

With these procedures performed, the solar energy potential of a site can be estimated. Extensive local measurements are necessary in order to evaluate the utilization potential of the solar radiation energy in a particular region.

Solar installation sites must be carefully selected. The primary energy supply and the presumed energy demand are the decisive factors in determining the economic feasibility of a particular site. Local measurements should include the following quantities:

- global radiation G
- direct solar radiation S
- diffuse sky radiation H
- number of hours of sunshine SD
- degree of cloudiness N
- air temperature T_A
- wind direction D
- wind intensity F

If possible, the measurement should be conducted over a relatively long period (several years). The transformation of the sun energy to electricity by photovoltaic panels and heat energy by solar thermal collectors depends on the type and model of collectors. Therefore, the efficiency of the chosen collector should be considered for the calculation of site solar energy potential.

5.4.2. Exploitation of Solar Energy

Today, solar energy is being utilized in many ways at various scales. On a small scale, it is used at the household level in goods such as watches, cookers and heaters. The medium-scale uses, such as in solar architecture houses, include water heating and irrigation. At the community level, it can be used for water pumping, water desalination, purification and rural electrification. On an industrial scale, solar energy is used for power generation, detoxification, municipal water heating and telecommunication. In general, there are actually two basic ways to use solar energy.

5.4.3. Solar Thermal System

While heat from the sun (over $300°C$) is utilized on a large scale in electricity generation, it can also be used in small- to medium-scale heating, cooling, cooking and drying equipment. Solar thermo-electric technologies utilize energy from the sun in the form of heat to generate electricity. The sun evaporates a fluid from which heat transfer systems may be used to operate an engine that drives a power conversion system. In a solar thermo-electric system, sunlight is concentrated with mirrors or lenses to attain a high temperature sufficient for power generation. Parabolic trough systems, central-receiver systems, parabolic dish systems and solar ponds are among those used.

The basic components of a solar thermo–electric system (Figure 5.5) are a collector system (which is the panels that collect the solar radiation), a receiver system, a transport storage system (mainly in the form of fluid that transfers the heat between the systems) and lastly a power conversion system converting energy from one form to another.

Solar water heaters are relatively simple solar thermal applications that transform solar radiation into heat that is used to warm water for heating, washing, cooking and cleaning. Solar water heaters consist of glass-covered collectors with a dark colored or especially coated absorber panel inside. Water (used as the heat transfer fluid) is warmed by the sun and can be stored in insulated tanks for later use. There are two main components of a typical solar water heating system: the flat plate solar collector and the hot water storage tank. Flat plate collectors absorb the solar radiation and conduct the heat to water that circulates through the collector in pipes.

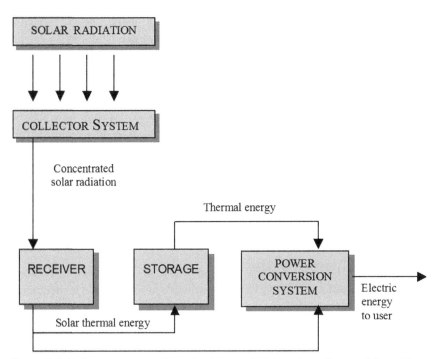

Figure 5.5 *Main components of a solar thermo-electric system.* *(Source: Adapted from De Laquil et al., 1993).*

Table 5.5 Various applications of solar thermal systems in an agricultural farm

Temperature range	Applications
Low grade thermal energy < 100 °C	Water heating, air heating, space heating, space cooling, communication, etc.
Medium grade thermal energy 100–300 °C	Cooking, drying, pumping, irrigation, water desalination

Another solar thermal application is the solar drier. There are often two stages of the process: first, solar radiation is captured and used to heat air; then comes the actual drying during which heated air moves through, warms and extracts moisture from the product. Drying takes place in a large box called the drying chamber. Air is either heated in a flat plate collector or directly via a window in the drying chamber.

All these components of the solar thermal system could be used on a farm and contribute by minimizing the costs of energy supply (see Table 5.5).

5.4.4. Solar Photovoltaic

Solar cells are produced from wafers of silicon (a form of pure sand) which is chemically treated and then arranged in parallel or in series in a module/panel. It can also be a film coating that is applied to a glass plate. The photovoltaic effect occurs when light falls on an active photovoltaic surface. This energy penetrates the cell near the junction between the p-type and n-type silicon. The semiconductor dislodges an electron, leaving behind a "hole." The electrons generated in this process of electron-hole pair formation tend to migrate to the n-region in the front contacts and can flow into an external circuit (see Figure 5.6).

Important applications of PV-systems in agriculture include:

- Electrification (lighting for buildings, power supply to remote locations)
- Solar pumps for water pumping
- Household and office appliances (ventilation, air conditioners, computers, emergency power, battery chargers, etc.)
- Communication (PV-powered remote radio telephones or repeaters)
- Solar desalination

The main components of a PV system (Figure 5.7) include a module, a battery, a battery control unit/charge controller, a DC-AC inverter (where necessary) and the load (appliances).

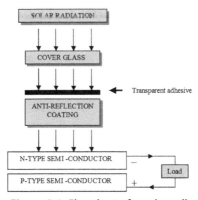

Figure 5.6 *Flowchart of a solar cell.*

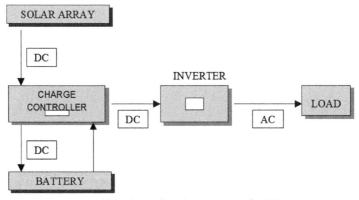

Figure 5.7 *Flowchart of main structure of a PV system.*

5.5. DATA COLLECTION AND PROCESSING FOR ENERGY UTILIZATION

The following basic data on solar radiation and climate conditions are essential for the development of preliminary designs for the solar plants:

- average global radiation density Q_i **(kW/m²)**
- average direct radiation density Q_1 **(kW/m²)**
- average duration of sunshine S **(h/d or h/a)**
- average ambient temperature T_u **(°C)**

The local global radiation density Q_i, the duration of sunshine, the temperature T_u and indicators for the direct solar radiation Q_1, should be obtained from the records of national measuring stations. The annual curves of both the average daily and monthly values, as well as the average annual values, are relevant information.

In addition, at the site of a projected solar installation, measurements should be taken of the global radiation level, the direct solar radiation level, the ambient temperature and the atmospheric humidity. From the humidity data general conclusions can be drawn as to the climatic conditions as well as the variations exhibited by Q_i, Q_1, T_u and S.

5.5.1. Water and Space Heating

For water and space heating, the potential consumer must first define the energy requirement Q_v. It is calculated as follows:

$$Q_v = m_v \cdot C_1 \cdot (T_2 - T_1)(kJ/h) \tag{5.5}$$

where m_v = quantity of water to be heated per hour (kg/h), C_1 = specific heat of water = 4.18 (kJ /kg deg. [°C]), T_1, T_2 = initial temperature of the cold water, final temperature of the warm water (°C).

The energy supply Q_s of a solar installation can be calculated as follows:

$$Q_s = \vartheta \cdot Q_i \cdot A \cdot \cos\alpha(kJ/h) \tag{5.6}$$

where ϑ = overall efficiency of the solar plant; Qi = average global radiation density (kJ/m² h); A = collector area of the solar installation (m²); α = angle between a line perpendicular to the surface and the direction of radiation.

By combining equations (5.5) and (5.6) the required collector area A can be obtained as follows:

$$A = m_v \cdot c_1 / \vartheta \cdot Q_i \cdot \cos\alpha (T_2 - T_1)(m^2) \tag{5.7}$$

5.5.2. Drying of Agricultural Produce

The energy requirement Q_v of solar drying plants is a function of the following factors:
- type of material to be dried
- moisture content of the material to be dried before and after drying
- maximum permissible drying temperature
- temperature and relative humidity of the ambient air
- drying time

Therefore the necessary collector area is calculated as follows:

$$A = Q_v / \vartheta \cdot Q_i \cdot \cos\alpha(m^2)$$

5.6. WIND ENERGY

The bandwidth of the power density of wind is very large. The value of the ten-minute maximum is many times higher than the annual average. This broad variation in the power density of wind causes specific challenges for the choice and construction of wind power generators.

BOX 5.12 The Annual Value of Wind Energy Is Calculated as Follows:

$$e_a = \sigma_L/2 \int V^3 dt (J/m^2)$$

e_a = specific annual energy (J/m^2)
σ_L = air density (kg/m^3)
V = speed of wind (m/s)
t_1 = beginning of year
t_2 = end of the year

BOX 5.13 The Total Efficiency of a Wind Generator Consists of Three Components:

$\vartheta = C_p\, \vartheta_m\, \vartheta_{AM}$
C_p = performance coefficient
ϑ_m = mechanical efficiency
ϑ_{AM} = efficiency of linked-up machines

The performance coefficient C_p is a measure for the aerodynamic quality of a windmill. The mechanical efficiency ϑ_m registers the losses in the transmission bearings and the gear. The transformation losses in the linked-up machine, e.g. generator or pumps, are considered by the efficiency ϑ_{AM}. So the efficiency of a wind energy converter is the ratio of the specific efficiency to the maximum theoretical wind potential. Modern wind generators achieve a total efficiency ϑ of 40% or more.

The successful exploitation of wind energy is site specific depending on the wind resources of the area being exploited. The economic viability of wind energy converters depends on the wind conditions that prevail at a particular site. Electricity generation from wind energy requires generally a wind speed higher than 4 meters per second (m/s). For wind pumps, lower wind speeds can be sufficient. However, most wind pumps will not start below a wind speed of 3 m/s and will furl at about 12 to 15 m/s.

From a technical point of view, we can install the wind converter in favorable wind fields at distances of 5 to 8 rotor diameters. At this distance, the direct mutual influencing of the wind converters is still small; this gives a surface factor—defined as proportion of rotor face to field area—of 0.016 or less.

For the installation of wind converters, in addition to the existence of reasonable wind speed on the site, there are other limiting factors to be considered. For example:
- use limitation and prohibitions in:
  - settlements and traffic areas
  - forest areas
  - natural and landscape conservation areas
  - leisure and recreation areas
- required safety distances and limitations from aesthetic aspects.

These restrictions affect turbines of various sizes differently.

On the other hand, the agricultural land use is not limited by windmill installations if the turbines are not too small and are not placed too densely. The area required for foundations does not hinder the activities of fodder or crop production.

5.7. BIOMASS

Biomass includes all materials of organic origin (e.g., all natural living or growing materials and their residues). The delimitation compared to fossil energy carriers begins with peat, a secondary product of rotting organic matter. Therefore, all plants and animals, their residues and wastes as well as materials resulting from their transformation (paper-cellulose), organic wastes from the food industry, as well as organic wastes of households and industrial production, qualify as biomass.

Biomass appears in different forms, which is simultaneously produced in organisms (see Table 5.6). Cellulose is the most frequent organic substance.

Table 5.6 Different forms of biomass and their worldwide annual growth

Biomass form	Worldwide annual growth	
	%	Billion t/y
Cellulose	65	100
Hemicellulose	17	27
Lignin	17	27
Starch		1
Sugar	1	0.1
Fat		
Protein		0.13
Dyes		
Sum	**100**	**155**

Cellulose is a polysaccharide, consisting of pure glucose chains, which have been connected by hydrogen in crystal bonds.

Woody plants consist of 20 to 30% hemicelluloses. They are also polysaccharides but consist not only of pure glucose chains but other sugars as well.

The wood pulp lignin constitutes about 30% of the woody plants. Lignin causes the lignification of vegetable cells by occlusion into the cellulose matrix. Compared with cellulose and lignin the remaining other forms of biomass play a small role.

Starch (1050 million t/y), sugar (100 million t/y) as well as fat, protein and dyes (130 million t/y) constitute only 1% of the world biomass production.

Biomass represents one of the most important renewable energy resources for the future. It is used diversely as an energy carrier (Table 5.7).

In thermo-chemical procedures, biomass is transferred to secondary energy carriers by oxidation, by application of heat or by chemical processing. In the agricultural sector the biomass can be used as an energy resource by combustion, or as fuel and biogas.

5.7.1. Energetic Use of Biomass
5.7.1.1. Combustion
The oldest thermo-chemical use of biomass is combustion. Almost half of the present forest woodcutting in the world is exploited as an energy resource for heating and cooking. Table 5.8 shows the portion of firewood

Table 5.7 Exploitation of biomass as an energy resource

Produced from	Wastes		Renewable raw materials		
	dry	moist			
Form of biomass	Wood, straw	Liquid manure, harvest residues	Sugar beet, cereals, maize	Oil plants	Cereals, C4-plants, small woods
Transformation techniques	Combustion Gasification, pyrolysis	Anaerobic fermentation (digestion)	Fermentation	Extraction	Combustion Pyrolysis
Products (Energy carriers)	Pyrolysis oil gas	Methane (biogas)	Alcohol	Veg. oil (methylester)	Pyrolysis oil, gas
Exploitation field	Heat, fuel, electricity				

Table 5.8 The part of firewood of total wood occurrence

	Wood (Total) million m^3	Firewood million m^3	Part of Firewood in %
Industrialized countries	1446	235	16
North America	673	93	14
Western Europe	275	40	14
Eastern Europe	61	12	20
Oceania	33	3	9
Former USSR	355	81	23
Other	48	7.5	17
Developing countries	1983	1594	80.4
Africa	481	437	91
Latin America	409	293	72
Near East	57	42	74
Far East	1027	815	79
Other	9	5.8	64

consumption of total wood occurrence in industrial and developing countries. In the developing countries, people also use other biomass resources such as dung or agricultural waste (straw) as combustible materials.

The net energy per unit mass of matter freed during combustion is called the net heating value H_U. For biomass the heating value depends on the specific net heating value of the dry matter (DM), organic dry matter (ODM), and their part $(1 - x)$ of the total mass. It also depends on the specific evaporation heat of water.

The net heating value H_U represents the basis for the further calculation of the energetic use of biomass in an Integrated Energy Farm.

The energetic value of biomass as a fuel depends on its humidity content. Rather than using biomass with a high humidity content for power generation, other procedures are preferable, such as biogas generation.

BOX 5.14 Net Energy Per Unit Mass of Matter Freed During Combustion

$$H_U = (1 - x)H_{UTS} - x2,441 (MJ/kg)$$

H_{UTS} = heating value of biomass dry matter

x = moist

2,441 MJ/kg = energy of initial temperature of 25°C

Table 5.9 Dry matter heating value of different types of biomass

Combustible biomass	Heating value H_{uts} of TS MJ/kg
Ash (tree)	18.6
Beech	18.8
Oak	18.3
Coniferous woods	19.0
Paper wastes	17.0
Straw	16.0
Sugar cane	15.0
Leaves	18.0
Fats	39.0
Fuel oil	43.0

(Kleemann, M., Meliß, M. 1988).

Table 5.8 shows that the net heating value of leaves and coniferous woods does not vary much between the different types. The heating value depends more on the moisture content than the type of wood (Table 5.9).

For absolutely dry wood the combustion temperature reaches approximately 1200°C. It decreases with increasing humidity content, simultaneously increasing the consumption of wood per energy unit produced.

Wood efficiencies vary significantly depending upon the type of application and combustion stove.

- open light 5–10%
- simple stove 20–30%
- open chimney 10–30%
- stove, cooking stove 40–50%
- fuel burnout stove (20–400 kW) 60–70%
- sub-fire stove (20–1200 kW) 60–80%

5.7.1.2. Extraction

Extraction is the second type of physical bioconversion (Figure 5.8). Direct extraction entails the separation of energy carriers from the biomass, for example through cold or hot pressing, steam breaking, acid hydrolysis or other procedures. Some plants are able to produce, in addition to the partly oxidized C-H-combinations as cellulose or lignin, oxygen-free hydrocarbons, which can be used directly as vegetable oils or as energy carriers. However, vegetable oils are employed today mainly for food production as well as for the production of enamels, colors, soaps and cosmetic items. The use of vegetable oils for heating and drying as well as for engines is limited by region. An extension of the cultivable areas can lead in some countries to

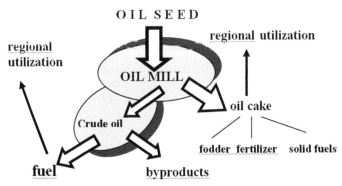

Figure 5.8 *Oil extraction, products, byproducts and utilization. (Source: Krause 1995).*

a competitive situation with food production. Above all, as in the developing countries where a food deficit exists, the effort to utilize vegetable oil as fuel can be problematic. Here the principle "food over fuel" should be retained.

Under specific conditions, however, in the developing countries, it is regionally possible to set up adapted techniques for extraction of vegetable oil for food production and for use in engines in order to strive for a largely autonomous, regional economy that is in cycle with nature.

5.7.2. Biogas Production

While thermal gasification requires biomass having less humidity and alcoholic fermentation depends on bio raw materials and residues containing sugar and starch, for biogas generation we need liquid or semi-liquid raw materials and residues. Here the material composition does not have a great importance. Solid biomass must only have small lignin content and a sufficiently large surface.

During the biogas generation, different bacteria groups are in interplay with each other and transform organic material under anaerobic conditions. Biogas is a mixture of methane and carbon dioxide.

The formation of methane from organic material occurs in three steps. The first two steps are the preparation steps; the third step is the process of methane production. In each step of the conversion process there are other bacteria influencing the process. These bacteria multiply themselves during the conversion.

The three steps of the process are:

1. Acid formation or hydrolysis
2. Acid disassembly
3. Methane production

Table 5.10 Biogas production from different biowastes and residues

Part 1

Branch	Waste type	VS [%]	CH₄ production per VS [Nm³/kg]	Biogas production per ton biomass [Nm³/ton]
Agriculture	Milking cows	8.5	0.21	27
	Young stock	7.3	0.21	23
	One year sows	6.4	0.29	28
	Porkers	5.8	0.29	24
	Hens	31	0.29	136
	Broilers	45	0.29	195
Pig slaughterhouse	Content of stomach/gut	16–20	0.46	110–138
	Fat and flotation slurry	4.5–35	0.50	34–263
	Remains after sieving	12	0.30	54

Part 2

Branch	Waste type	VS [%]	CH₄ production per VS [Nm³/kg]	Biogas production per ton biomass [Nm³/ton]
Cattle slaughterhouse	Content of stomach/gut	16	0.40	96
	Fat and flotation slurry	36	0.58	313
	Remains after sieving	12	0.30	54
Poultry slaughterhouse	Fat and flotation slurry	7–40	0.61	84–366
Dairies	Flotation slurry	7–8	0.40	42–48
	Whey	4–6	0.33	20–30
Oil mills	Bleaching clay	40	1.00	600
	Misc. organic material	25	0.47	176
Margarine	Fat slurry	90	0.81	1094

(Continued)

Table 5.10 Biogas production from different biowastes and residues—cont'd

Part 2

Branch	Waste type	VS [%]	CH$_4$ production per VS [Nm^3/kg]	Biogas production per ton biomass [Nm^3/ton]
Potato flour	Fruit sap	4	0.35	21
Pectin industry	Kelp remains	4–5	0.21	13–16
Brewery	Yeast/dregs	10	0.26	39
	Yeast/dregs	21	0.26	82
	Remains from filters	11	0.26	43
Pharmacy industry	Production slurry	5–100	0.30	23–450
Tannery	Glue leather	17	0.50	128

Part 3

Branch	Waste type	VS [%]	CH$_4$ production per VS [Nm^3/kg]	Biogas production per ton biomass [Nm^3/ton]
Vegetable market	Sap from vegetables	2.5–5	0.45	17–34
Fish oil/meal industry	Fat and flotation slurry	8–24	0.36	43–136
Fish filleting industry	Fat and flotation slurry	7–20	0.45	47–135
Herring cannery	Miscellaneous slurry	8–11	0.55	66–91
Mackerel cannery	Fat and flotation slurry	17–23	0.55	140–190
Shellfish industry	Fat and flotation slurry	20–26	0.75	225–231
Smoke fish industry	Fat and flotation slurry	8–44	0.59	71–389

Source: folkecenter for renewable energy (2000).

The quantity of gas produced as well as the methane content essentially depends on the following influencing parameters that determine the environment for bacteria:

- type of substrate
- dry matter
- temperature
- retention time
- pH-value
- quantity of substrate

Basic materials for methane extraction can be liquid manure, plant waste, and byproducts of food production (Table 5.10). All of them contain decomposable materials such as proteins, fat or carbohydrates (starch or cellulose).

The optimal dry matter content of the substrate to be fermented is 5 to 12%. This value can be adjusted by addition of water or urine. An optimal gas production is achieved in a pH- field from 6.5 to 7.2. In general the biogas plants work continuously. A specific quantity of biomass is added daily and a corresponding quantity of fermented substrate is diverted.

Liquid manure such as from large-scale animal husbandry is an excellent substrate for biogas generation. In contrast, solid manure (with straw mixtures) often leads to considerable technical difficulties. In Tables 5.11, 5.12 and 5.13 are presented some data referring to biogas generation.

Decreasing the space load increases the fermentation time. The optimal duration of fermentation, such as for mesophilic bacteria, is between 20 to 30 days. For example, if we have a substrate density of 1000 kg /m^3 and

---

BOX 5.15 An Important Factor for a Biogas Plant Working Continuously Is the Space Load (R_b). It Specifies How Much (kg) Organic Dry Matter May Be Loaded Per Day in a Cubic Meter of Fermenter Volume.

$$R_b = m_{su} C_{OTM} / V_R [(kg/d)/m^3]$$

m_{su} : daily supply with substrate (kg/d)

C_{OTM} : concentration of organic dry matter

V_R : reactor volume (m^3)

BOX 5.16 When the Reactor Is Filled, We Have Approximately the Following Situation:

$$V_R = V_{SU}(m^3)$$

V_{SU}: volume of substrate (m^3)

BOX 5.17 Where C_{OTM} Is:

$$C_{OTM} = m_{OTM}/m_{su} = V_{OTM}\mu_{OTM}/V_{SU}\mu_{SU}$$

μ: density (kg/ m^3)
m: mass (kg)
V: volume (m^3

BOX 5.18 The Fermentation Time of the Substrate t_{vw} in Days Can Be Calculated as Follows:

$$t_{vw} = V_{su}/V_{su} = m_{su}/m_{su}$$

V_{su} : Volume of daily supply of substrate (m^3 / d)

Table 5.11 Composition of animal excrement, referring to dry matter

	Pig excrement in %	Cattle excrement in %	Chicken excrement in %
Carbohydrates	48	20	25
Fat	4	4	4
Protein	19	15	29
Crude fiber	20	40	15
Ash	19	21	27

Table 5.12 Quantity of animal excreta per day

Quantity	Cattle per unit	Per CU	Pig per unit	Per CU = 5 units	Chicken per unit	Per CU = 250 units
Liquid manure (kg/d)	50	50	4	20	0,1	25
Dry matter (kg/ODM/d)	5	5	0,4	2	0,03	7,5
Quantity (1gas/kg ODM)	300	300	400	400	400	400
Quantity (1gas/d)	1500	1500	160	800	12	3000
Methane content in %	60	60	70	70	70	70
Heating value (Mj/Nm^3 Gas)	20	20	23	23	23	23

CU = cattle unit = 500 kg life weight
ODM = Organic Dry Matter

Table 5.13 Quantity of gas and retention time of agricultural products

Material	Biogas m^3 /kg ODM	Retention time (d)
Wheat straw	0.367	78
Sugar beet leaves	0.501	14
Potato tops	0.606	53
Maize tops	0.514	52
Clover	0.445	28
Grass	0.557	25

a concentration of 0.08, we will have a daily dry matter supply of 2.7 to 3 kg per m^3 fermenter volume corresponding to the quantity of 33.3 to 37.5 kg substrate. The cumulative gas production of animal excrement conducts at 30 °C is 0.380 to 0.400 m^3 of biogas per kg ODM. That corresponds to a daily gas gain per m^3 fermenter volume of 1.026 to 1.2 m^3. The production conditions are strongly influenced by the type and the composition of the substrate.

REFERENCES

De Laquil, P., Kearney, D., Geyer, M., Diver, M., 1993. In: Johansson, T.B., Kelly, H., Reddy, A.K.N., Williams, R.H., Burham, L. (Eds.), 'Solar thermal electric technology', Renewable Energy Sources for Fuels and Electricity. Island Press, Washington D.C.

El Bassam, N., Maegaard, P., 2004. Integrated Renewable Energy for Rural Communities. In: Planning Guidelines, Technologies and Applications (1st ed.). Elsevier Science, published June 30, pp. 14–49.

FAO (Farm and Agriculture Organization), 1995, Environment and Natural Resources Working Paper. http://www.fao.org/DOCREP/003/X8054E/X8054E00.HTM.

Maegaard P., http://www.folkecenter.net/gb/overview/.

Stout, B.A., 1990. Handbook of energy for world agriculture. Elsevier Applied Sciences, London.

CHAPTER SIX

Energy Basics, Resources, Global Contribution and Applications

Renewable energy is energy that comes from natural resources such as sunlight, wind, rain, tides, and geothermal heat, which are renewable (naturally replenished).

Renewable energy flows involve natural phenomena such as sunlight, wind, tides, plant growth, and geothermal heat. As the International Energy Agency (IEA) explains:

> *Renewable energy is derived from natural processes that are replenished constantly. In its various forms, it derives directly from the sun, or from heat generated deep within the earth. Included in the definition are electricity and heat generated from solar, wind, ocean, hydropower, biomass, geothermal resources, and biofuels and hydrogen derived from renewable resources. (IEA Renewable Energy Working Party 2002).*

6.1. BASICS OF ENERGY

Understanding some energy basics helps to work through what may be possible for a community, how renewable energy works and why its use should be considered.

Energy can be defined as "the ability to do work" and is measured in joules (J). The rate at which energy is generated or used is measured in watts. One watt (W) is one joule per second (1 J/s). The unit of watts most commonly used when discussing energy consumption is the kilowatt: i.e., 1000 watts, or 1 kW.

6.1.1. Energy Rating

Electrical appliances are rated in kilowatts. So, for example, an oil-filled radiant heater is rated at 1.5 kW. This means that when the heater is switched on it will immediately consume up to a maximum of 1.5 kW. Where large amounts of energy are generated or consumed, the units used are more likely to be in one of the following formats: megawatt

(1,000,000 watts or 1 MW), gigawatt (1,000,000,000 watts or 1 GW), or even terawatt (1,000,000,000,000 watts or 1 TW).

6.1.2. Energy Consumption

Units of energy consumption are usually expressed in terms of the amount of energy used over a certain period; the standard term for this is kilowatt hours or kWh—that is, the amount of energy consumed over an hour. A 1.5 kW heater, if left on for an hour with a constant electrical supply, will therefore consume 1.5 kWh of energy. By the same token, a 60 W lightbulb left on for an hour will consume 0.06 kWh = 60 watts × 1 hour = 60 watt hours or 0.06 kWh. Electricity is sold by the kWh, which equals 1 unit. The current domestic tariff is around $.065 per kWh. Therefore keeping the electric heater on for 1 hour will consume 1.5 units of electricity—$.10.

6.1.3. Energy Generation

The same rationale is applied to energy generation. Generators are rated in kW or MW, indicating the maximum that can be generated at any moment. If a 1 kW generator is operating at full capacity for 1 hour, it will generate 1 kWh.

However, the amount of energy generated will depend on how much useful energy is available to power the generator. It will only generate to its maximum rated level if it is supplied with sufficient useful energy. This applies equally to a small diesel generator or a wind generator; the only difference is that a small diesel generator will generally either be full on (with fuel), or off (no fuel), whereas the output from a wind generator will vary with wind speed.

6.2. GLOBAL CONTRIBUTION

About 16% of global final energy consumption comes from renewables, with 10% coming from traditional biomass, which is mainly used for heating, and 3.4% from hydroelectricity. New renewables (small hydro, modern biomass, wind, solar, geothermal, and biofuels) accounted for another 3% and are growing very rapidly. The share of renewables in electricity generation is around 19%, with 16% of global electricity coming from hydroelectricity and 3% from new renewables. The graph in Figure 6.1 shows global power capacities of various renewables, excluding hydro, from 2004 to 2010.

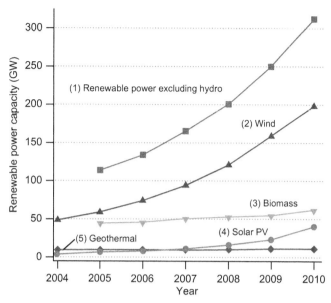

Figure 6.1 *Global renewable power capacities (excluding hydro). (Renewable Energy Policy Network for the 21st Century [REN21] 2006–2011).* (See color plate 9.)

Wind power is growing at the rate of 30% annually, with a worldwide installed capacity of 198 gigawatts (GW) in 2010 and is widely used in Europe, Asia, and the United States. At the end of 2011 the photovoltaic (PV) capacity worldwide was 67.4 GW. End 2011 PV was, after hydro and wind power, the third most important renewable energy in terms of installed capacity. The new grid-connected PV capacities were 16.6 GW in 2010 and 27.7 GW in 2011. In 2011 the top cumulative installed capacity, including the PV power stations, were in (GW): Germany 24.7, Italy 12.5, Japan 4.7, Spain 4.2, the USA 4.2, China 2.9, France 2.5, Belgium 1.5 and Australia 1.2. Solar thermal power stations operate in the USA and Spain, and the largest of these is the 354 megawatt (MW) SEGS power plant in the Mojave Desert. The world's largest geothermal power installation is the Geysers in California, with a rated capacity of 750 MW. Brazil has one of the largest renewable energy programs in the world, involving production of ethanol fuel from sugarcane, and ethanol now provides 18% of the country's automotive fuel. Ethanol fuel is also widely available in the USA.

While many renewable energy projects are large-scale, renewable technologies are also suited to rural and remote areas, where energy is often crucial in human development. As of 2011, small solar PV systems provide electricity to a few million households, and micro-hydro configured into mini grids serves many more. Over 44 million households use biogas made in household-scale digesters for lighting and/or cooking and more than 166 million households rely on a new generation of more efficient biomass cookstoves. United Nations' Secretary-General Ban Ki-moon has said that renewable energy has the ability to lift the poorest nations to new levels of prosperity.

Climate change concerns, coupled with high oil prices, peak oil, and increasing government support, are driving increasing renewable energy legislation, incentives and commercialization. New government spending, regulation and policies helped the industry weather the global financial crisis better than many other sectors. According to a 2011 projection by the International Energy Agency, solar power generators may produce most of the world's electricity within 50 years, dramatically reducing the emissions of greenhouse gases that harm the environment.

6.3. RESOURCES AND APPLICATIONS

Renewable energy replaces conventional fuels in four distinct areas: electricity generation, hot water/space heating, motor fuels (REN21 2010):

- **Power generation**. Renewable energy provides 19% of electricity generation worldwide. Renewable power generators are spread across many countries, and wind power alone already provides a significant share of electricity in some areas: for example, 14% in the U.S. state of Iowa, 40% in the northern German state of Schleswig-Holstein, and 20% in Denmark. Some countries get most of their power from renewables, including Iceland and Paraguay (100%), Norway (98%), Brazil (86%), Austria (62%), New Zealand (65%), and Sweden (54%).

- **Heating**. Solar hot water makes an important contribution to renewable heat in many countries, most notably in China, which now has 70% of the global total (180 GWth). Most of these systems are installed on multi-family apartment buildings and meet a portion of the hot water needs of an estimated 50–60 million households in China. Worldwide, total installed solar water heating systems meet a portion of the water heating needs of over 70 million households. The use of biomass for heating continues to grow as well. In Sweden, national use of biomass energy has surpassed that of oil. Direct geothermal for heating is also growing rapidly.

- **Transport fuels.** Renewable biofuels have contributed to a significant decline in oil consumption in the United States since 2006. The 93 billion liters (about 24.6 billion US gallons) of biofuels produced worldwide in 2009 displaced the equivalent of an estimated 68 billion liters (about 18 billion US gallons) of gasoline, equal to about 5% of world gasoline production (REN21 2010).

Figure 6.2 compares the applications and output of the major sources of renewable energy.

Source	Utilization	Output
Sunlight - heat	Solar water heating	Hot water
Biomass - wood	Combustion - boiler or stove	Heat
Sunlight - heat from sun transferred to soil, air or water	Ground source heat pump Air source heat pump Water source heat pump Passive solar	Heat and hot water
Sunlight	Solar photovoltaic cells (PV)	Electricity
Wind	Wind turbine	Electricity
Water	Hydro turbine	Electricity
Biomass - wood	Combustion - boiler (+ steam turbine if electricity generation is desired)	Heat (and electricity)
Biomass - biodegradable matter	Anaerobic digestion (decomposition without oxygen, producing methane gas) - can also use the gas to generate electricity if desired	Heat (and electricity)
Wave (wind)	Floating or shore based electrical generators converting kinetic energy from waves	Electricity
Tidal	Underwater electrical generators converting kinetic energy from tides	Electricity

Figure 6.2 *Main sources of renewable energy. (Community Energy Scotland, 2011).*

REFERENCES

Community Energy Scotland, 2011. Community renewable energy toolkit [WWW]. Available from: http://www.scotland.gov.uk/Resource/Doc/264789/0079289.pdf.
Renewable Energy Policy Network for the 21st Century (REN21), 2006–2011. Renewables Global Status Report [WWW], Available from: http://www.ren21.net/REN21Activities/Publications/GlobalStatusReport/tabid/5434/Default.aspx.
Clean Edge, 2009. Clean Energy Trends, 1–4.
European Photovoltaic Industry Association, 2012. Market Report 2011.
Global Energy Wind Council (GWEC), Press release Brussels, February 2, 2007. Global wind energy markets continue to boom – 2006 another record year (PDF).
IEA Renewable Energy Working Party, 2002. Renewable Energy. into the mainstream p. 9.
Kroldrup, Lars, February 15, 2010. Gains in Global Wind Capacity Reported Green Inc.
Leone, Steve, August 25 2011. U.N. Secretary-General: Renewables Can End Energy Poverty. Renewable Energy World.

Morales, Alex, February 07, 2012. Wind Power Market Rose to 41 Gigawatts in 2011, Led by China. Bloomberg.

REN21, 2010. Renewables 2010 Global Status Report, p. 15, 53.

REN21, 2011. Renewables 2011: Global Status Report, p. 14, 15, 17, 18.

Renewableenergyaccess.com. America and Brazil Intersect on Ethanol; Retrieved 11-21-2011.

Sills, Ben, Aug 29, 2011. Solar May Produce Most of World's Power by 2060, IEA Says. Bloomberg.

Solar Trough Power Plants (PDF) http://www.osti.gov/accomplishments/documents/fullText/ACC0196.pdf.

United Nations Environment Program Global Trends in Sustainable Energy Investment, 2007. Analysis of Trends and Issues in the Financing of Renewable Energy and Energy Efficiency in OECD and Developing Countries (PDF), p. 3.

World Energy Assessment, 2001. Renewable energy technologies, p. 221.

CHAPTER SEVEN

Solar Energy

Electromagnetic energy (solar radiation) transmitted by the sun (approximately one billionth of which reaches the earth) is the basis of all terrestrial life. "It amounts to about 420 trillion kilowatt-hours, and is several thousand times greater than all the energy used by all people. Solar energy is harnessed by capturing the sun's heat (through solar heaters) or light (through photovoltaic cells). It is estimated that one square kilometer (about 0.4 square miles) of land area receives some 4000 kilowatts (4 megawatts) of solar energy every day, enough for the requirements of a medium-sized town" (Business Dictionary 2012).

Solar technologies are broadly characterized as either passive solar or active solar depending on the way they capture, convert and distribute solar energy. Active solar techniques include the use of photovoltaic panels and solar thermal collectors to harness the energy. Passive solar techniques include orienting a building to the sun, selecting materials with favorable thermal mass or light dispersing properties, and designing spaces that naturally circulate air.

Solar-powered electrical generation relies on heat engines and photovoltaics. Solar energy's uses are limited only by human ingenuity. A partial list of solar applications includes space heating and cooling through solar architecture, potable water via distillation and disinfection, daylighting, solar hot water, solar cooking, and high temperature process heat for industrial purposes. To harvest the solar energy, the most common method is to use solar panels (TheFreeDictionary.com 2012).

Solar power is the conversion of sunlight into electricity. Sunlight can be converted directly into electricity using photovoltaics (PV), or indirectly with concentrated solar power (CSP), which normally focuses the sun's energy to boil water, which is then used to provide power. Other technologies also exist, such as Stirling engine dishes, which use a Stirling cycle engine to power a generator.

7.1. PHOTOVOLTAIC

Photovoltaics were initially used to power small- and medium-sized applications, from the calculator powered by a single solar cell to off-grid homes powered by a photovoltaic array.

Photovoltaics convert light into electric current using the photoelectric effect.

A photovoltaic system (or PV system) is a system that uses one or more solar panels to convert sunlight into electricity. It consists of multiple components, including the photovoltaic modules, mechanical and electrical connections and mountings and means of regulating or modifying the electrical output.

A solar cell, or photovoltaic cell (PV), is a device that converts light into electric current using the photoelectric effect. The first solar cell was constructed by Charles Fritts (Perlin 1999) in the 1880s. In 1931 a German engineer, Dr. Bruno Lange (Popular Science 1931), developed a photo cell using silver selenide in place of copper oxide. Although the prototype selenium cells converted less than 1% of incident light into electricity, both Ernst Werner von Siemens and James Clerk Maxwell recognized the importance of this discovery. Following the work of Russell Ohl in the 1940s, researchers Gerald Pearson, Calvin Fuller and Daryl Chapin created the silicon solar cell in 1954 (Perlin 1999).These early solar cells cost 286 USD/watt and reached efficiencies of 4.5–6% (Perlin 1999).

Different materials display different efficiencies and have different costs. Materials for efficient solar cells must have characteristics matched to the spectrum of available light. Some cells are designed to efficiently convert wavelengths of solar light that reach the Earth's surface. However, some solar cells are optimized for light absorption beyond Earth's atmosphere as well. Light-absorbing materials can often be used in *multiple physical configurations* to take advantage of different light absorption and charge separation mechanisms.

Materials presently used for photovoltaic solar cells include mono-crystalline silicon, polycrystalline silicon, amorphous silicon, cadmium telluride, and copper indium selenide/sulfide (Jacobson 2009).

Photovoltaic modules and arrays produce direct-current (DC) electricity. They can be connected in both series and parallel electrical arrangements to produce any required voltage and current combination (NASA 2002).

Many currently available solar cells are made from bulk materials that are cut into wafers between 180 to 240 micrometers thick and are then processed like other semiconductors.

Other materials are made as thin-films layers, organic dyes, and organic polymers that are deposited on supporting substrates. A third group are made from nanocrystals and used as quantum dots (electron-confined

(a)

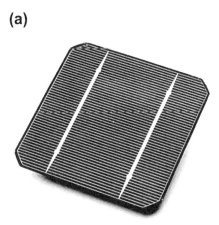

Figure 7.1a *A solar cell made from a monocrystalline silicon wafer and polycrystalline photovoltaic cells Laminated to backing material in a module.* *(Department of Energy http://www.eere.energy.gov/solar/pv_systems.html, Retrieved Aug. 17, 2005).*

(b)

Figure 7.1b *Multiple modules can be wired together to form an array; in general, the larger the area of a module or array, the more electricity produced.* *(Georg Slickers, 2006). http://en.wikipedia.org/wiki/Creative_Commons.*

nanoparticles). Silicon remains the only material that is well-researched in both *bulk* and *thin-film* forms.

As mentioned, solar cells produce direct current (DC) power, which fluctuates with the intensity of the irradiated light. This usually requires conversion to certain desired voltages or alternating current (AC), which requires the use of inverters. Multiple solar cells are connected inside the modules. Modules are wired together to form arrays, then tied to an inverter, which produces power with the desired voltage and frequency/ phase (when it's AC).

Trackers and sensors to optimize performance are often seen as optional, but tracking systems can increase viable output by up to 100% (RISE 2010). PV arrays that approach or exceed one megawatt often use solar trackers. Accounting for clouds, and the fact that most of the world is not on the equator, and that the sun sets in the evening, the correct measure of solar power is *insolation*—the average number of kilowatt-hours per square meter per day. For the weather and latitudes of the United States and Europe, typical insolation ranges from $4kWh/m^2/day$ in northern climes to $6.5kWh/m^2/day$ in the sunniest regions (Whitlock 2000).

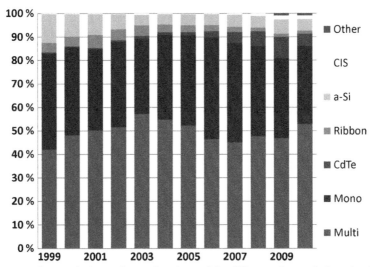

Figure 7.2 *The graph shows the market share of the different photovoltaic technologies from 1999 Until 2011. The light and dark blue are multi- and mono crystalline silicon, respectively; together they represented 87% of the market in 2010.* (Cleanenergy 2011). http://en.wikipedia.org/wiki/File:PV_Technology.png. (See color plate 10.)

For large systems, the energy gained by using tracking systems outweighs the added complexity (trackers can increase efficiency by 30% or more) (Utility Scale Solar Power Plants (PDF), 2011).

7.1.2. Applications

Most commercially available solar panels are capable of producing electricity for at least 20 years (Zweibel, K., 2010). The typical warranty given by panel manufacturers is over 90% of rated output for the first 10 years, and over 80% for the second 10 years. Panels are expected to function for a period of 30 to 35 years (Shenzhen JCN New Energy Technology CO).

Many residential systems are connected to the grid wherever available, especially in the developed countries with large markets. In these grid-connected PV systems, use of energy storages is optional (IEA-PVPS 2011). In certain applications such as satellites, lighthouses, or in developing countries, batteries or additional power generators are often added as back-ups, which form stand-alone power systems.

Between 1970 and 1983 photovoltaic installations grew rapidly, but falling oil prices in the early 1980s moderated the growth of PV from 1984 to 1996. Since 1997, PV development has accelerated due to supply issues with oil and natural gas, global warming concerns, and the improving economic position of PV relative to other energy technologies (Solar: photovoltaic 2009). Photovoltaic production growth has averaged 40% per year since 2000 and installed capacity reached 39.8 GW at the end of 2010, with 17.4 GW of that total in Germany (BP Statistical World Energy Review 2011). As of October 2011, the largest photovoltaic (PV) power plants in the world are the Sarnia Photovoltaic Power Plant (Canada, 97 MW), Montalto di Castro Photovoltaic Power Station (Italy, 84.2 MW) and Finsterwalde Solar Park (Germany, 80.7 MW) (PV Resources.com 2011)

The view (Figure 7.6a) of the International Space Station (ISS) was taken while it was docked with the Space Shuttle Atlantis and shows parts of all but one of the current components. From the top are the Progress supply vehicle, the Zvezda service module, and the Zarya functional cargo block (FGB) (NASA Science/Science News 2011).

7.2. CONCENTRATING SOLAR THERMAL POWER (CSP)

Concentrated solar power (CSP) systems use lenses or mirrors and tracking systems to focus a large area of sunlight into a small beam.

Figure 7.3 *Photovoltaic (PV) power plant. (Dennis Schroeder, NREL/PIX 19176).*

Figure 7.4 *Waldpolenz Solar Park, Germany. First Solar 40-MW CDTe PV Array installed by JUWI Group in Waldpolenz, Germany. (JUWI Solar GmbH, Energie-Allee 1, 55286 Wörrstadt, Germany, 2008). http://www.juwi.com/solar_energy/large_scale_plants. html.*

Commercial concentrated solar power plants were first developed in the 1980s. The 354 MW SEGS CSP installation is the largest solar power plant in the world, located in the Mojave Desert of California. Other large CSP plants include the Solnova solar power station (150 MW) and the Andasol solar power station (150 MW), both in Spain (Solar Millennium AG 2011). The 200 MW Golmud Solar Park in China is the world's largest photovoltaic plant (Wang, Ucilia 2011). Many power plants today use fossil fuels as a heat source to boil water. The steam from the boiling water spins a large

Figure 7.5 *The 71.8MW Lieberose Photovoltaic Park in Germany. (JUWI Group 2008). http://www.juwi.com/solar_energy/references/lieberose_solar_park.html.*

(a)

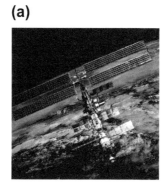

Figure 7.6a *Solar panels on the International Space Station absorb light from both sides. These Bifacial cells are more efficient and operate at lower temperature than singlesided equivalents. (NASA Science/Science News 2011).* (See color plate 11.)

(b)

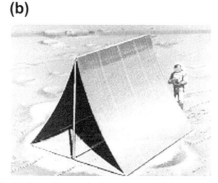

Figure 7.6b *A self-deploying photovoltaic array on the moon. Artist's concept by Les Bossinas, NASA Lewis Research Center. (NASA Science/Science News 2011).* (See color plate 12.)

Figure 7.7 *A camel transports cooling box for medicine, powered by solar energy in the Sahara. (Solar Power Panels, 2009). http://solarpowerpanels.ws/solar-power/camels-carry-solar-powered-refrigerators-for-mobile-health-clinics.*

turbine, which drives a generator to produce electricity. However, a new generation of power plants with concentrating solar power systems uses the sun as a heat source. The three main types of concentrating solar power systems are: *linear concentrator, dish/engine,* and *power tower systems.*

Linear concentrator systems collect the sun's energy using long rectangular, curved (U-shaped) mirrors. The mirrors are tilted toward the sun, focusing sunlight on tubes (or receivers) that run the length of the mirrors. The reflected sunlight heats a fluid flowing through the tubes. The hot fluid then is used to boil water in a conventional steam-turbine generator to produce electricity. There are two major types of linear concentrator systems: parabolic trough systems, where receiver tubes are positioned along the focal line of each parabolic mirror; and linear Fresnel reflector systems, where one receiver tube is positioned above several mirrors to allow the mirrors greater mobility in tracking the sun.

A dish/engine system uses a mirrored dish similar to a very large satellite dish. The dish-shaped surface directs and concentrates sunlight onto a thermal receiver, which absorbs and collects the heat and transfers it to the engine generator. The most common type of heat engine used today in dish/engine systems is the Stirling engine. This system uses the fluid heated by the receiver to move pistons and create mechanical power. The mechanical power is then used to run a generator or alternator to produce electricity.

A power tower system uses a large field of flat, sun-tracking mirrors known as heliostats to focus and concentrate sunlight onto a receiver on the

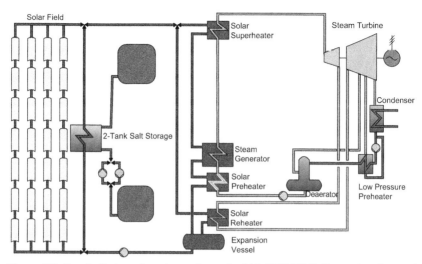

Figure 7.8 *Parabolic trough system schematic. (U.S. DOE 2001).* (See color plate 13.)

top of a tower. A heat–transfer fluid heated in the receiver is used to generate steam, which, in turn, is used in a conventional turbine generator to produce electricity. Some power towers use water/steam as the heat–transfer fluid. Other advanced designs are experimenting with molten nitrate salt because of its superior heat–transfer and energy-storage capabilities. The energy-storage capability, or thermal storage, allows the system to continue to dispatch electricity during cloudy weather or at night.

Figure 7.9 *64 MW Nevada Solar 1 solar plant. (U.S. DOE 2001).* (See color plate 14.)

Figure 7.10 *SEGS trough plants.* *(U.S. DOE 2001).* (See color plate 15.)

Figure 7.11 *Nevada Solar 1 CSP collector.* *(U.S. DOE 2001).* (See color plate 16.)

7.3. SOLAR THERMAL COLLECTORS

Solar thermal collectors transform solar radiation into heat and transfer that heat to a medium (water, solar fluid, or air). Solar water heating (SWH) or solar hot water (SHW) systems have been well established for many years, and are widely used throughout the world. In a "close-coupled" SWH system the storage tank is horizontally mounted directly above the solar collectors on the roof. No pumping is required as the hot water naturally

Figure 7.12 *Linear Fresnel Collector (Ausra). (U.S. DOE 2001).*

Figure 7.13 *A commercial unit under development by Abengoa called PS10, an 11 MW plant in Seville, Spain. (Photo: Abengoa Solar).*

rises into the tank through passive heat exchange. In a "pump-circulated" system the storage tank is ground or floor mounted below the level of the collectors; a circulating pump moves the water or heat transfer fluid between

Figure 7.14 *Prototype 150 kW dish/Stirling power plant at Sandia National Laboratory. (U.S. DOE 2001).*

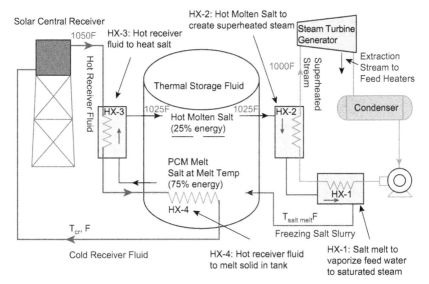

Figure 7.15 *Heat transfer and latent heat Storage in inorganic molten salts for concentrating solar power plants. (Stekli 2010).*

the tank and the collectors. There are multiple types of solar thermal collectors:

- **Evacuated tube collectors** are the most efficient but most costly type of hot water solar collectors. These collectors have glass or metal tubes with a vacuum, allowing them to operate well in colder climates.
- **Batch solar water heaters**, also called integral collector–storage systems, have storage tanks or tubes inside an insulated box, the south side of which is glazed to capture the sun's energy.
- **Flat plate collector**, a box covered by glass or plastic with a metal absorber plate on the bottom. The glazing, or coating, on the absorber plate helps to better absorb and retain heat.

- **Unglazed flat-plate collectors**, typically made from rubber, are primarily used for heating pools.
- **Air collectors** are used primarily for space heating in the home. Flat-plate solar collectors—durable, weatherproof boxes that contain a dark absorber plate located under a transparent cover—are still the most common type of collector used for water heating in many countries despite being inferior to evacuated tube collectors in many ways.

Evacuated heat pipe tubes are designed such that convection and heat loss are eliminated, whereas flat-plate solar panels contain an air gap between absorber and cover plate that allows heat loss to occur. Further, thermal heat pipe systems are capable of limiting the maximum working temperature, whereas flat-plate systems have no internal method of limiting heat build-up, which can cause system failure. Finally, evacuated heat pipe systems are lightweight, easy to install and require minimal maintenance. Flat-plate systems, on the other hand, are difficult to install and maintain, and must be completely replaced should one part of the system stop working.

Figures 7.16 and 7.17 show two types of solar collectors that are commonly being installed in South Australia.

An **evacuated tube solar collector** is composed of hollow glass tubes. All the air is removed from the tubes to create a vacuum that acts as an excellent insulator. An absorber coating inside the tube absorbs the solar radiation. This energy is transferred to the fluid moving through the collector and then to the hot water storage tank. In cooler climates a heat exchanger is used to separate the potable water from the non-toxic anti-freeze in the collector.

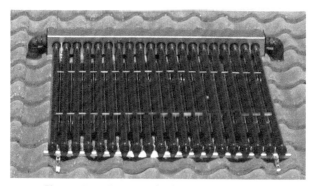

Figure 7.16 *Evacuated tube.* (Hot Water Now).

Figure 7.17 *Flat-plate collectors.* *(Solar Tribune 2012).*

Batch solar water heaters, also called integral collector–storage (ICS) systems, are made up of a water tank or tubes inside an insulated, glazed box. Cold water flows through the solar collector. The water is heated and then continues on to the backup water heating storage tank. Some water can be stored in the collector until it is needed. ICS systems are a type of direct solar water heating system, which circulates water to be heated, rather than using a heat transfer fluid to capture the solar radiation.

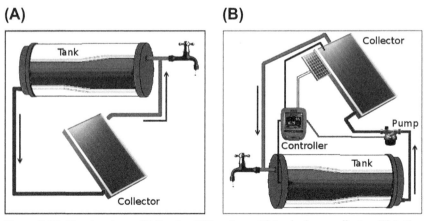

Figure 7.18 *Direct systems: (A) Passive CHS system with tank above collector. (B) Active system with pump and controller driven by a photovoltaic panel. (Jwhferguson, self-published work 2010, accessed from URL http://www.solarcontact.com/solar-water/heater).* (See color plate 17.)

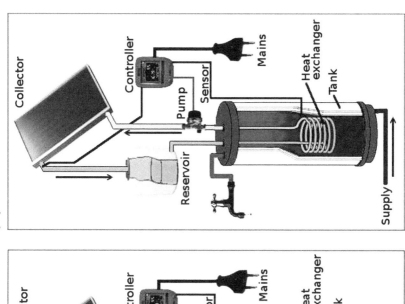

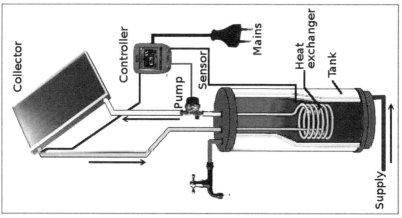

Figure 7.19 *Indirect active systems: (C) Indirect system with heat exchanger in tank. (D) Drainback system with drainback reservoir. In these schematics the controller and pump are driven by mains electricity.* Euro-Flachstecker_2.jpg: SomnusDe 2010, Wolff Mechanical Inc, accessed from URL http://azairconditioning.com/residential/solar-heaters/. (See color plate 18.)

A **flat plate solar collector** is an insulated box covered by glass or plastic with a metal absorber plate on the bottom. The weatherproofed collectors are usually glazed with a coating to better absorb and retain heat. Heat transfer fluid flows through metal tubes lying below the absorber plate. The fluid then flows through a heat exchanger before entering the storage tank. **Unglazed flat-plate collectors** (without insulation or absorber coatings) do not operate in cool or windy climates but are excellent for heating water in a pool (Solar Tribune 2012).

Solar hot **air collectors** are mounted on south-facing vertical walls or roofs. Solar radiation reaching the collector heats the absorber plate. Air passing through the collector picks up heat from the absorber plate.

Freezing, overheating and leaks are less troublesome for solar air collectors than for liquid collectors. But since liquid is a better heat conductor, solar collectors using water or a heat transfer fluid are more suited to hot water heating for the home. A solar hot air collector is most often used for space heating. There are two types of air collectors: glazed and unglazed (Energy4You 2012).

SWH systems are designed to deliver hot water for most of the year. In colder climates a gas or electric booster may be needed as a backup to deliver sufficient hot water.

Figure 7.20 *Bolivia Inti-Sud Soleil solar cooker construction workshop.* http://solarcooking.wikia.com/wiki/solar_cookers_world_net.

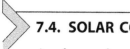

7.4. SOLAR COOKERS AND SOLAR OVENS

A **solar cooker, or solar oven** is a device to generate heat from the energy of solar radiation. The radiation is concentrated in the focus of a concave mirror. The solar cookers are suitable for

- Boiling water for heating food
- Cooking, roasting, baking, grilling or frying
- Juice production
- Small food stalls in the open air
- Commercial use in dyeing
- Soap production
- Processing of natural fibers for weaving

Box types are constructed from metal or mirror reflectors and a glass plate. Aligned with the sun, the light is reflected by the reflector through the glass into the interior of the furnace. At about 150 ° C, water will boil to cook food. The parabolic types consist of mirrors focusing sunlight on a mostly matte black container in the focus area. The container absorbs solar radiation, so that its contents are heated.

Because they use no fuel and cost nothing to operate, many non-profit organizations are promoting their use worldwide to help reduce fuel costs for low-income people, reduce air pollution and slow deforestation and desertification, caused by use of firewood for cooking.

Many types of cookers exist. Simple solar cookers use the following basic principles:

- Concentrating sunlight: A reflective mirror of polished glass, metal or metallized film is used to concentrate light and heat from the sun into a small cooking area, making the energy more concentrated and increasing its heating power.
- Converting light to heat: A black or low reflectivity surface on a food container or the inside of a solar cooker will improve the effectiveness of turning light into heat. Light absorption converts the sun's visible light into heat, substantially improving the effectiveness of the cooker.
- Trapping heat: It is important to reduce convection by isolating the air inside the cooker from the air outside the cooker. A plastic bag or tightly sealed glass cover will trap the hot air inside. This makes it possible to reach similar temperatures on cold and windy days as on hot days.
- Greenhouse effect: Glass transmits visible light but blocks infrared thermal radiation from escaping. This amplifies the heat trapping effect (Solar Cooker Designs 2011).

REFERENCES

"BP Statistical World Energy Review 2011" (XLS). Retrieved 8 August 2011.

Business Dictionary, 2012. Solar energy [WWW], Available from: http://www. businessdictionary.com/definition/solar-energy.html#ixzz1lsoyXA8S.

Energy4You, 2012. Solar hot air [WWW]. Available from: http://www.energy4you.net/ hotair.htm.

Energy Society of Canada Inc. and Solar, Oct. 21-24, 2000.

Hot Water Now [n.d.] Hot Water Adelaide, Solar, Gas & Heat Pump [WWW]. Available from www.hotwaternow.com.au/colchester_solar_panel.jpg.

Jacobson, Mark Z, 2009. Review of Solutions to Global Warming Air Pollution, and Energy Security 4.

NASA Science/Science News, 2002. How do photovoltaics work [WWW]. Available from: http://science.nasa.gov/science-news/science-at-nasa/2002/solarcells/.

NASA Science/Science News, 2011. The edge of sunshine [WWW]. Available from: http://science.nasa.gov/science-news/science-at-nasa/2002/08jan_sunshine/.

Perlin, John, 1999. From Space to Earth (The Story of Solar Electricity). Harvard University Press, ISBN 0-674-01013-2, p. 147.

Perlin, p. 29–30, 38.

Perlin, p. 18–20.

Popular Science, June 1931. Magic Plates, Tap Sun For Power, pp. 41,134.

PV Resources.com, 2011. World's largest photovoltaic power plants.

Rise & Shine 2000, the 26th Annual Conference of the Solar, Halifax, Nova Scotia, Canada.

Shenzhen JCN New Energy Technology CO., Ltd http://www.jcnsolar.com/help/help-0001, 0002, 0015.shtml.

"Small Photovoltaic Arrays". Research Institute for Sustainable Energy (RISE), Murdoch University. Retrieved 5 February 2010. (Modified by El Bassam).

"Solar Cooker Designs". Solar Cookers World Network. 2011 http://solarcooking.wikia. com/wiki/Solar_Cookers_World_Network_%28Home%29.

Solar Millennium AG, 2011. http://www.solarmillennium.de/english/technology/ references_and_projects/andasol-spain/index.html.

Solar: photovoltaic: Lighting Up The World. Retrieved 19 May 2009.

Solar Tribune, 2012. Guide to solar thermal energy [WWW] Available from: http:// solartribune.com/solar-thermal-power/#.T0_PvfU0iuI.

Stekli, J., 2010. DOE CSP R&D: Storage Award Overview. U.S. Department of Energy: Energy Efficiency & Renewable Energy: Solar Energy Technologies [WWW]. Available from DOE CSP R&D: Storage Award Overview, DOE HQ | April 28, 2010.

TheFreeDictionary.com, 2012. Solar energy [WWW]. Available from. http://encyclopedia. thefreedictionary.com/solar+energy.

"Trends in Photovoltaic Applications Survey report of selected IEA countries between 1992 and 2009, IEA-PVPS". Retrieved 8 November 2011.

U.S. Department of Energy (DOE), 2001. Concentrating solar power commercial application study: Reducing water consumption of concentrating solar power electricity generation A report to Congress. Available from: http://www1.eere.energy.gov/solar/ pdfs/csp_water_study.pdf.

U.S. Department of Energy (DOE) EERE, 2011. Photovoltaics [WWW]. Available from: http://www1.eere.energy.gov/solar/pdfs/52481.pdf.

Utility Scale Solar Power Plants, (PDF), 2011.

Wang, Ucilia, 6 June 2011. The Rise of Concentrating Solar Thermal Power. Renewable Energy World.

Whitlock, C. E., et al. Release 3 NASA Surface Meteorology and Solar Energy Data Set for Renewable Energy Industry Use.

Zweibel, K., 2010. Should solar photovoltaics be deployed sooner because of long operating life at low, predictable cost? Energy Policy. http://dx.doi.org/10.1016/j.enpol.2010.07.040.

CHAPTER EIGHT

Wind Energy

Conversion of wind power into wind energy has been utilized for many years. Some examples are:

- Wind turbines to make electricity
- Windmills for mechanical power
- Wind pumps for drinking, irrigation or drainage
- Sails to propel ships

A windmill converts the energy in the wind into electrical energy or mechanical energy to pump water or grind cereals. The most common windmills in operation today generate power from three-blade, horizontal-axis windmills with the nacelle mounted on steel towers that can be cylindrical steel plate or lattice towers. This modern windmill concept has emerged since 1977 and has become the industrial standard. The nacelles of horizontal-axis windmills usually include a gearbox, asynchronous generator, and other supporting mechanical and electrical equipment.

The total amount of economically extractable power available from the wind is considerably more than present human power use from all sources. At the end of 2010, worldwide nameplate capacity of wind-powered generators was 197 gigawatts (GW). Wind power now has the capacity to generate 430 TWh annually, which is about 2.5% of worldwide electricity usage (World Wind Energy Report 2010). Over the past five years the average annual growth in new installations has been 27.6 percent. Wind power market penetration is expected to reach 3.35 percent by 2013 and 8 percent by 2018 (Renewable Energy World 2009.) Several countries have already achieved relatively high levels of wind power penetration, such as 21% of stationary electricity production in Denmark (Danish Energy Agency 2012), 18% in Portugal (Monthly Statistics – SEN 2012), 16% in Spain (The Spanish electricity system 2011), 14% in Ireland (Renewables 2010) and 9% in Germany in 2010. As of 2011, 83 countries around the world are using wind power on a commercial basis (REN21 2011). A large wind farm may consist of several hundred individual wind turbines which are connected to the electric power transmission network. Offshore wind power can harness the better wind speeds that are available offshore compared to on land, so offshore wind power's contribution in terms of electricity supplied is higher. Small onshore wind facilities are used to provide electricity to isolated locations and utility companies increasingly buy back surplus electricity produced by small domestic wind turbines. Although a variable source of power, the intermittency of wind seldom creates problems when using wind power to supply up to 20% of total electricity demand, but as the proportion rises, increased costs, a need to use storage such as pumped-storage hydroelectricity, upgrade the grid, or a lowered ability to supplant conventional production may occur. Power management techniques such as excess capacity, storage, dispatchable backing supply (usually natural gas), exporting and importing power to neighboring areas or reducing demand when wind production is low, can mitigate these problems.

Wind power, as an alternative to fossil fuels, is plentiful, renewable, widely distributed, clean, produces no greenhouse gas emissions during operation, and uses little land (Fthenakis 2009). In operation, the overall cost per unit of energy produced is similar to the cost for new coal and natural gas installations. The construction of wind farms (Figures 8.1) is not universally welcomed, but any effects on the environment from wind power are generally much less problematic than those of any other power source.

A wind turbine is a device that converts kinetic energy from the wind into mechanical energy. If the mechanical energy is used to produce electricity, the device may be called a wind generator or wind charger. If the mechanical energy is used to drive machinery, such as for grinding grain or pumping water, the device is called a windmill or wind pump. Developed for over a millennium, today's wind turbines are manufactured in a range of vertical and horizontal ax is types. The smallest turbines are used for

Figure 8.1 *Lillgrund Wind Farm's wind turbines in the Sound near Copenhagen and Malmö. (Mariusz Paździora).* (See color plate 19.)

applications such as battery charging or auxiliary power on sailing boats, while large grid-connected arrays of turbines are becoming an increasingly large source of commercial electric power (Hau 2006).

8.1. GLOBAL MARKET

The world market for wind turbines set a new record in the year 2011 and reached a total size of 42 gigawatts, after 37.6 gigawatts in 2010. The total capacity worldwide has come close to 239 gigawatts, enough to cover 3 % of the world's electricity demand.

Among the individual countries, China kept its strong position reaching a similar amount to the previous year, 2010; China installed around 18 GW of new wind turbines during 2011, coming to a total capacity of 63 GW, more than one-fourth of the global wind capacity. The second largest market for new wind turbines was again the USA with 6.8 GW, followed by India (2.7 GW), Germany (2 GW) and a surprisingly strong Canada with 1.3 GW of new installed capacity. Spain, France and Italy each added around 1 GW (WWEA 2012).

A strong increase in wind power utilization can be observed especially in the emerging markets, like China, India, Brazil, and Mexico. This opens new windows for further growth, as these countries do have an increasing

Country	Total Capacity end of 2011 [MW]	Added Capacity 2011[MW]	Total Capacity end 2010 [MW]	Added Capacity 2010 [MW]	Total Capacity end 2009 [MW]
China*	62.733	18.000	44.733	18.928	25.810
Usa	46.919	6.810	40.180	5.600	35.159
Germany	29.075	2.007	27.215	1.551	25.777
Spain	21.673	1.050	20.676	1.515	18.865
India*	15.800	2.700	13.065	1.258	11.807
Italy*	6.747	950	5.797	950	4.850
France	6.640	980	5.660	1.086	4.574
United Kingdom	6.018	730	5.203	962	4.245
Canada	5.265	1.267	4.008	690	3.319
Portungal*	4.290	588	3.702	345	3.357
Denmark	3.927	180	3.803	309	3.460
Sweden	2.816	746	2.052	603	1.450
Japan	2.501	167	2.334	251	2.083
Rest of the World*	24.200	6.000	18.201	3.191	15.010
Total*	238.604	42.175	196.629	37.642	159.756

*-Preliminary Data   ©WWEA 2012

Figure 8.2 *Capacity of wind turbines worldwide. (WWEA 2012).*

need for electricity, which can be matched by wind power in a very economic, safe and timely way. On the other hand, several of the European markets showed stagnation or even a decrease. The US market presented itself stronger than in 2010; however, the mid-term prospects are not very bright, due to a lack of clarity regarding the political support schemes. Canada now finds itself as number five in terms of new capacity, aided by Ontario with its Green Energy Act, adopted as a consequence of the WWEC2008 (WWEA 2012).

8.2. TYPES OF WIND TURBINES

Wind turbines can rotate about either a horizontal or a vertical axis, the former being both older and more common.

8.2.1. Horizontal-axis Wind Turbines

Horizontal-axis wind turbines (HAWT) have the main rotor shaft and electrical generator at the top of a tower, and must be pointed into the wind. Small turbines are pointed by a simple wind vane, while large turbines generally use a wind sensor coupled with a servo motor. Most have a gearbox, which turns the slow rotation of the blades into a quicker rotation that is more suitable to drive an electrical generator (see http://www.windpower.org/en/tour/wtrb/comp/index.htm).

Since a tower produces turbulence behind it, the turbine is usually positioned upwind of its supporting tower. Turbine blades are made stiff to prevent the blades from being pushed into the tower by high winds. Additionally, the blades are placed a considerable distance in front of the tower and are sometimes tilted forward into the wind a small amount.

Turbines used in wind farms for commercial production of electric power are usually three-bladed and pointed into the wind by computer-controlled motors. These have high tip speeds of over 320 km/h (200 mph), high efficiency, and low torque ripple, which contribute to good reliability. The blades are usually colored light gray to blend in with the clouds and range in length from 20 to 40 meters (66 to 130 ft) or more. The tubular steel towers range from 60 to 90 meters (200 to 300 ft) tall. The blades rotate at 10 to 22 revolutions per minute. At 22 rotations per minute the tip speed exceeds 90 meters per second (300 ft/s) (AWEO.org). A gear box is commonly used for stepping up the speed of the generator, although designs may also use direct drive of an annular generator. Some models operate at constant speed, but more energy can be collected by variable-speed turbines, which use a solid-state power converter to interface to the transmission system. All turbines are equipped with protective features to avoid damage at high wind speeds, by feathering the blades into the wind, which ceases their rotation, supplemented by brakes.

8.2.2. Vertical-axis Design

Vertical-axis wind turbines (or VAWTs) have the main rotor shaft arranged vertically. Key advantages of this arrangement are that the turbine does not need to be pointed into the wind to be effective. This is an advantage on sites where the wind direction is highly variable, for example when integrated into buildings. The key disadvantages include the low rotational speed with the consequential higher torque and hence higher cost of the drive train, the inherently lower power coefficient, the 360-degree rotation of the airfoil within the wind flow during each cycle and hence the highly dynamic loading on the blade, the pulsating torque generated by some rotor designs on the drive train, and the difficulty of modeling the wind flow accurately and hence the challenges of analyzing and designing the rotor prior to fabricating a prototype (see http://www.awsopenwind.org/downloads/documentation/ModelingUncertaintyPublic.pdf).

With a vertical axis, the generator and gearbox can be placed near the ground, using a direct drive from the rotor assembly to the ground-based gearbox, hence improving accessibility for maintenance.

When turbines are mounted on a rooftop, the building generally redirects the wind over the roof and this can double the wind speed of the turbine. If the height of the rooftop mounted turbine tower is approximately 50% of the building height, this is near the optimum for maximum wind energy and minimum wind turbulence. It should be kept in mind that wind speeds within the built environment are generally much lower than at exposed rural sites (see http://www.urbanwind.net/pdf/technological_analysis.pdf).

Another type of vertical axis is the parallel turbine; similar to the cross flow fan or centrifugal fan it uses the ground effect. Vertical-axis turbines of this type have been tried for many years (Brill 2002). The Magenn Wind Kite blimp uses this configuration as well, chosen because of the ease of running (Insource/Outsource 2007).

Subtypes of the vertical-axis design include those described in the following sections.

8.2.2.1. Darrieus Wind Turbine

"Eggbeater" turbines, or Darrieus turbines, were named after the French inventor, Georges Darrieus. (see http://www.symscape.com/blog/vertical_axis_wind_turbine) They have good efficiency, but produce large torque ripple and cyclical stress on the tower, which contributes to poor reliability. They also generally require some external power source, or an additional Savonius rotor to start turning, because the starting torque is very low. The torque ripple is reduced by using three or more blades, which results in greater solidity of the rotor. Solidity is measured by blade area divided by the rotor area. Newer Darrieus type turbines are not held up by guy-wires but have an external superstructure connected to the top bearing (Singh, 2012)

8.2.2.2. Giromill

A subtype of the Darrieus turbine has straight, as opposed to curved, blades. The cyclo-turbine variety has variable pitch to reduce the torque pulsation and is self-starting. (see http://www.awea.org/faq/vawt.html) The advantages of variable pitch are: high starting torque; a wide, relatively flat torque curve; a lower blade speed ratio; a higher coefficient of performance; more efficient operation in turbulent winds; and a lower blade-speed ratio which lowers blade bending stresses. Straight, V, or curved blades may be used (see Experimental Mechanics 1978, Volume 18, Number 1 – SpringerLink).

8.2.2.3. Savonius Wind Turbine

These are drag-type devices with two (or more) scoops that are used in anemometers, *Flettner* vents (commonly seen on bus and van roofs), and in some high-reliability low-efficiency power turbines. They are always self-starting if there are at least three scoops (see http://www.flettner.co.uk/home.htm).

8.2.2.4. Twisted Savonius

Twisted Savonius is a modified Savonius, with long helical scoops to give a smooth torque; this is mostly used as a roof wind turbine or on some boats (such as the Hornblower Hybrid) (Varnon 2010).

8.3. SMALL WIND TURBINES

Small wind turbines are electric generators that utilize wind energy to produce clean, emissions-free power for individual homes, farms, and small businesses. With this simple and increasingly popular technology, individuals can generate their own power and cut their energy bills while helping to protect the environment. The U.S. leads the world in the production of small wind turbines, which are defined as having rated capacities of 100 kilowatts and less, and the market is expected to continue strong growth through the next decade.

The small wind turbines may be used for a variety of applications including on- or off-grid residences, telecom towers, offshore platforms, rural schools and clinics, remote monitoring and other purposes that require energy where there is no electric grid, or where the grid is unstable. They may be as small as a 50-watt generator for boat or caravan use. The U.S. Department of Energy's National Renewable Energy Laboratory (NREL) defines small wind turbines as those smaller than or equal to 100 kilowatts. Small units often have direct drive generators, direct current output, air-elastic blades, lifetime bearings and use a vane to point into the wind.

Larger, more costly turbines generally have geared power trains, alternating current output, flaps and are actively pointed into the wind. Direct drive generators and air-elastic blades for large wind turbines are being researched (see Small Wind, U.S. Department of Energy National Renewable Energy Laboratory website).

The American Wind Energy Association (AWEA) released a report in 2011on growth of the small wind turbine market.

Highlights of the report, as described by Zachary Shahan (2011), editor of the *Clean Technica* clean energy news website, include:

1. Small wind turbine market grew 26% in 2010 (in kW), more than any previous year.
2. Nearly 8,000 small wind power systems were sold in 2010 for a record $139 million.
3. Cumulative small wind turbine sales in the U.S. after 2010 bring U.S. capacity to 179 MW (from 144,000 units), nothing to laugh about.
4. Fewer wind turbines were sold, but they apparently had a higher average capacity. (Average cost was $5,430/kW.)
5. More of the sales were for grid-connected turbines than in previous years, representing 90% of sales for the first time ever.
6. Sales come from over a dozen small wind turbine companies, including over a half-dozen U.S. companies. "**Domestic sales** by U.S. manufacturers accounted for an 83% share of the U.S. market; on a unit basis, U.S. manufacturers claimed 94% of domestic sales."

Figure 8.3 *Vertical-axis wind turbine in Cap-Chat, Quebec. (Spiritrock4u 2008).* (See color plate 20.)

7. Turbines manufactured in the U.S. typically used 80% domestic content.
8. 51 different wind turbines models were known to be sold.

It seems clear from the AWEA report that small wind turbine sales and popularity are growing and that they are creating jobs for numerous Americans today (Shahan 2011).

8.4. GOOGLE SUPERHIGHWAY, USA

Google is continuing to invest in green power sources and has just invested in another wind farm, a project to build a 350-mile power transmission backbone along the U.S. East Coast.

Figure 8.4 *Offshore wind farm using 5MW turbines RE power 5M in the North Sea off Belgium.* *(Hillewaert 2008).*

The U.S. East Coast project will hopefully provide 60,000 MW of offshore wind potential. The area has relatively shallow waters that extend miles out to sea that makes it easier to erect wind turbines about 10–15 miles out of sight offshore.

"This system will act as a superhighway for clean energy. By putting-strong, secure transmission in place, the project removes a major barrier to scaling up offshore wind, an industry that despite its potential, only had its first federal lease signed last week and still has no operating projects in the U.S." (see http://www.compositesworld.com/news/google-donates-billions-for-us-offshore-wind-energy-transmission-upgrades).

Figure 8.5 *Eleven 7.5 MW E126 at Estinnes Windfarm, Belgium, two months before completion, with unique two-part blades.* *(Melipal1, July 2010).*

Figure 8.6 *Turbines at the Donghai Bridge wind farm, China's first offshore project.* *(Recharge 2012).* (See color plate 21.)

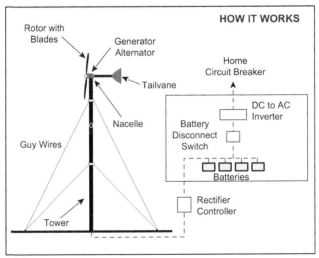

Figure 8.7 *Small wind energy device.* *http://www.ebay.com/itm/BUILD-WINDMILL-WIND-TURBINE-DIY-FREE-ENERGY-FOREVER-/200358886438.*

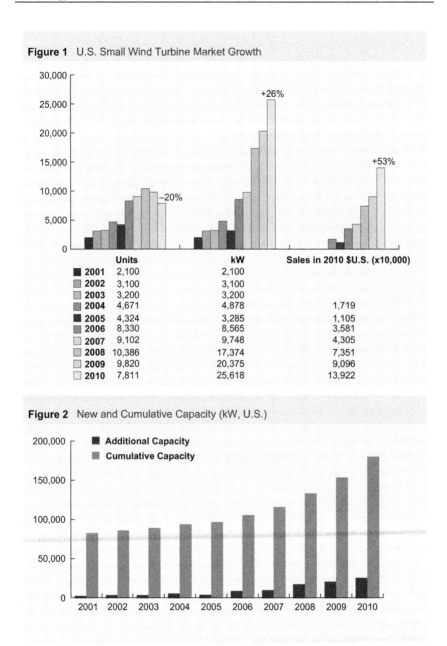

Figure 1 U.S. Small Wind Turbine Market Growth

	Units	kW	Sales in 2010 $U.S. (x10,000)
2001	2,100	2,100	
2002	3,100	3,100	
2003	3,200	3,200	
2004	4,671	4,878	1,719
2005	4,324	3,285	1,105
2006	8,330	8,565	3,581
2007	9,102	9,748	4,305
2008	10,386	17,374	7,351
2009	9,820	20,375	9,096
2010	7,811	25,618	13,922

Figure 2 New and Cumulative Capacity (kW, U.S.)

Figure 8.8 *(a) U.S. small wind turbine market growth. (b) New and cumulative capacity (kW, U.S.).* (See color plate 22.)

Figure 8.9 *Texas offshore wind farms, 2011. (World Future Energy Summit 2011).*

REFERENCES

American Wind Energy Association (AWEA), 2011. 2010 U.S. Small Wind Turbine Market Report [WWW]. Available from: http://www.awea.org/learnabout/smallwind/upload/2011_turbine_market_report.pdf.

BTM Forecasts 340-GW of Wind Energy by 2013. Renewableenergyworld.com 27, March 2009.

Exploit Nature-Renewable Energy Technologies by Gurmit Singh;, Aditya Books, p. 378.

Fthenakis, V., Kim, H.C., 2009. Land use and electricity generation: A life-cycle analysis. Renewable and Sustainable Energy Reviews 13 (6–7), 1465.

Hau, Erich, 2006. Wind turbines: fundamentals, technologies, application, economics Birkhäuser ISBN 3-540-24240-6.

Hillewaert,"http://commons.wikimedia.org/wiki/User:Biopics" \o "User:Biopics" Hans 2008, http://en.wikipedia.org/wiki/File:Windmills_D1-D4_%28Thornton_Bank%29.jpg.

Insource/Outsource: 2007-09-16, http://insourceoutsource.blogspot.com/2007_09_16_archive.html.

Månedlig elforsyningsstatistik, January 2012. HTML-spreadsheet summary tab B58-B72 Danish Energy Agency 18.

Marloff, Richard H. 1978, Stresses in turbine-blade tenons subjected to bending, Experimental Mechanics, Volume 18, Number 1 - SpringerLink (PDF).

Modular wind energy device – Brill, Bruce I. 2002, http://www.freepatentsonline.com/6481957.html.

"Monthly Statistics – SEN". February 2012.

Recharge, 2012. China continues offshore push with 300MW wind farm plan [WWW]. Available from: http://www.rechargenews.com/energy/wind/article297985.ece.

REN21, 2011. Renewables 2011: Global Status Report, p. 11.

Renewable and Sustainable Energy Reviews 13 (6–7): 1465.

Shahan, Z., 2011. Small wind turbine market growing strong in U.S [WWW]. Clean Technica. Available from: http://cleantechnica.com/2011/09/25/small-wind-turbine-market-growing-strong-in-u-s/.

"Renewables". eirgrid.com. Retrieved 22 November 2010.

Technical. Specs of Common Wind Turbine Models [AWEO.org].

"The Spanish electricity system: preliminary report 2011". January 2012. p. 13.

Varnon, Rob, December 2, 2010. Derecktor converting boat into hybrid passenger ferry. Connecticut Post website.

World Wind Energy Report 2010 (PDF). Report, February 2011. World Wind Energy Association.

Wikipedia, 2012b. Wind turbine [Online]. Available from: http://en.wikipedia.org/wiki/Wind_turbine#Types.

CHAPTER NINE

Biomass and Bioenergy

9.1. CHARACTERISTICS AND POTENTIALS

Of all renewable energy sources the largest contribution, especially in the short and medium range, is expected to come from biomass. Fuels derived from energy crops are not only potentially renewable, but are also sufficiently similar in origin to the fossil fuels (which also began as biomass) to provide direct substitution. They are storable, transportable, available and affordable and can be converted into a wide variety of energy carriers using existing and novel conversion technologies, and thus have the potential to be significant new sources of energy into the twenty-first century.

Biomass as energy feedstock is:

- Storable
- Transportable
- Convertible
- Available and affordable
- Always with positive energy balance

REN21 estimates the current bioenergy supply to be about 13.3 percent of the total primary energy demand of 50 EJ/year of which 7–10 EJ/year is used in industrial countries and 40–45 EJ/year is used in developing countries. China and India are the largest biomass energy producers worldwide. While most biomass electricity production occurs in OECD countries, several developing countries, especially India, Brazil, other Latin American/Caribbean and African countries, generate large amounts of electricity from combustion of bagasse from sugar alcohol production. Denmark, Finland, Sweden, and the Baltic countries provide substantial shares (5–50 percent) of district heating fuel. Among developing countries, small-scale power and heat production from agricultural waste is common, for example from rice or coconut husks. Biomass pellets have become more common, with about 6 million tons consumed in Europe in 2005, about half for residential heating and half for power generation (often in small-scale CHP plants). The main European countries employing pellets are Austria, Belgium, Denmark, Germany, Italy, the Netherlands, and Sweden. Although a global division of biomass consumption for heating versus power

Figure 9.1 *Our future oil fields: Transportation fuels will be extracted from non-food crops!* (El Bassam 2009). (See color plate 23.)

is not available, in Europe two thirds of biomass is used for heating. For the transport fuels sector, production of fuel ethanol for vehicles reached 39 billion liters in 2006, an 18 percent increase from 2005. Most of the increased production occurred in the United States, with significant increases also in Brazil, France, Germany, and Spain. The United States became the leading fuel ethanol producer in 2006, producing over 18 billion liters and jumping ahead of longstanding leader Brazil. US production increased by 20 percent as dozens of new production plants came on-line. Even so, production of ethanol in the United States could not keep up with demand during 2006, and ethanol imports increased sixfold, with about 2.3 billion liters imported in 2006. By 2007, most gasoline sold in the country was being blended with some share of ethanol. Biodiesel production jumped 50 percent in 2006, to over 6 billion liters globally. Half of world biodiesel production continued to be in Germany. Significant production increases also took place in Italy and the United States (where production more than tripled). In Europe, supported by new policies, biodiesel gained broader acceptance and market. The potential of bioenergy crops is huge. More than 450,000 plant species have been identified worldwide; approximately 3000 of these are used by humans as sources of food, energy and other feedstock. About 300 plant species have been domesticated as crops for agriculture; of these, 60 species are of major importance. It will be a vital task to increase the number of plant species that could be grown to produce fuel feedstock. This can be done by identifying, screening, adapting and breeding, and employing biotechnology and genetic engineering to a part of the large potential of the flora available on our planet. The introduction of new crops to agriculture would lead to an improvement in the biological and environmental condition of soils, water, vegetation and landscapes and increasing biodiversity. They can be converted by first-generation biofuels technologies (combustion, ethanol from sugar and starch crops, and biodiesel from oil crops) or the second-generation biofuels (hydrolysis or synthesis of lignocellulosic crops and their residues to liquid or gaseous fuels).

A large potential and opportunity could be expected from the production of fuels from microalgae extraction, which I would characterize as the third generation. They can produce at least ten times more biomass than land crops per unit area; they can be grown in saline, brackish or waste waters in ponds or in bioreactors and can be converted to oil, ethanol or hydrogen.

The potential contribution of bioenergy to the world energy demand of some 467 EJ per year (2004) may be increased considerably compared to the current 45–55 EJ. A range of 200–400 EJ per year in biomass harvested for energy production may be expected during this century. Assuming expected average conversion efficiencies, this would result in 130–260 EJ per year of transport fuels or 100–200 EJ per year of electricity.

9.2. SOLID BIOFUELS

There are four basic groups of plant species rich in lignin and cellulose that are suitable for conversion into solid biofuels (as bales, briquettes, pellets, chips, powder, etc.):
- Annual plant species such as cereals, pseudocereals, hemp, kenaf, maize, rapeseed, mustard, sunflower, and reed canary grass (whole plants)
- Perennial species harvested annually, such as *Miscanthus* and other reeds
- Fast-growing tree varieties like poplar, aspen or willow with a perennial harvest rhythm (short rotation or cutting cycle), short rotation coppice (SRC)
- Tree species with a long rotation cycle

Table 9.1 Overview of the global potential of biomass for energy (EJ per year) to 2050 for a number of categories and the main preconditions and assumptions that determine these potentials

Biomass Category	Main Assumptions and Remarks	Energy Potential in Biomass up to 2050
Energy farming on current agricultural lands	Potential land surplus: 0—4 Gha (average: 1—2 Gha). A large surplus requires structural adaptation toward more efficient agricultural production systems. When this is not feasible, the bioenergy potential could be reduced to zero. On average higher yields are likely because of better soil quality: 8—12 dry tonne/ha/yr★ is assumed.	0—700 EJ (more average development: 100—300 EJ)
Biomass production on marginal lands	On a global scale a maximum land surface of 1.7 Gha could be involved. Low productivity of 2—5 dry tonne/ha/yr.★ The net supplies could be low due to poor economics or competition with food production.	<60—110 EJ
Residues from agriculture	Potential depends on yield/product ratios and the total agricultural land area as well as type of production system. Extensive production systems require re-use of residues for maintaining soil fertility. Intensive systems allow for higher utilization rates of residues.	15—70 EJ
Forest residues	The sustainable energy potential of the world's forests is unclear — some natural forests are protected. Low value: includes limitations with respect to logistics and strict standards for removal of forest material. High value: technical potential. Figures include processing residues.	30—150 EJ
Dung	Use of dried dung. Low estimate based on global current use. High estimate: technical potential. Utilization (collection) in the longer term is uncertain.	5—55 EJ
Organic wastes	Estimate on basis of literature values. Strongly dependent on economic development, consumption and the use of bio-materials. Figures include the organic fraction of MSW and waste wood. Higher values possible by more intensive use of bio-materials.	5—50 EJ

Table 9.1 Overview of the global potential of biomass for energy (EJ per year) to 2050 for a number of categories and the main preconditions and assumptions that determine these potentials—cont'd

Biomass Category	Main Assumptions and Remarks	Energy Potential in Biomass up to 2050
Combined potential	Most pessimistic scenario: no land available for energy farming; only utilization of residues. Most optimistic scenario: intensive agriculture concentrated on the better quality soils. In parentheses: average potential in a world aiming for large-scale deployment of bioenergy.	40−1100 EJ (200−400 EJ)

*Heating value: 19 GJ/tonne dry matter.
(Berndes et al. 2003; Hoogwijk et al. 2005a; Smeets et al. 2007).

The raw materials can be used directly after mechanical treatment and compaction, or are converted to other types of biofuels. The lignocellulose plant species offer the greatest potential within the array of worldwide biomass feedstock. The following basic processes are of major importance for the conversion of lignocellulose biomass into fuel suitable for electricity production: direct combustion of biomass to produce high grade heat; advanced gasification to produce fuel gas of medium heating value; and flash pyrolysis to produce bio-oil, with the possibility of upgrading to give hydrocarbons similar to those in mineral crude oils. The production of methanol or hydrogen from woody biomass feedstock (such as lignocellulose biomass woodchips from fast growing trees, *Miscanthus* spp. or *Arundo donax*) via processes that begin with

Table 9.2 Projection of technical energy potential from power crops grown by 2050

	Africa	China	Latin America	Industrialized	All Regions
Available Area for Biomass Production in 2050 (Gha)	0.484	0.033	0.665	0.100	1.28
Maximum additional asset of Energy from Biomass (EJ/year)	145	21	200	30	396
Total (including traditional Biomass 45 EJ/year)				IPPC 2001	441

Source: (El Bassam, N 2004).

Table 9.3 Overview of current and projected performance data for the main conversion routes of biomass to power and heat and summary of technology status and deployment. Due to the variability of technological designs and conditions assumed, all costs are indicative.

Conversion Option	Typical Capacity	Net Efficiency (LHV basis)	Investment Cost Ranges (€/kW)	Status and Deployment
Biogas production via anaerobic digestion	Up to several MWe	10–15% electrical (assuming on-site production of electricity)		Well-established technology. Widely applied for homogeneous wet organic waste streams and waste water. To a lesser extent used for heterogeneous wet wastes such as organic domestic wastes.
Landfill gas production	Generally several hundred kWe	As above.		Very attractive GHG mitigation option. Widely applied and, in general, part of waste treatment policies of many countries.
Combustion for heat	Residential: 5–50 kWth Industrial: 1–5 MWth	Low for classic fireplaces, up to 70–90% for modern furnaces	100/kWth for logwood stoves, 300–800/ kWth for automatic furnaces, 300–700/ kWth for larger furnaces	Classic firewood use still widely deployed, but not growing. Replacement by modern heating systems (i.e., automated, flue gas cleaning, pellet firing) in e.g., Austria, Sweden, Germany ongoing for years.

Combined heat and power	0.1–1 MWe 1–20 MWe	60–90% (overall) 80–100% (overall)	3500 (Stirling) 2700 (ORC) 2500–3000 (Steam turbine)	Stirling engines, steam screw type engines, steam engines, and organic rankine cycle (ORC) processes are in demonstration for small-scale applications between 10 kW and 1 MWe. Steam turbine based systems 1–10 MWe are widely deployed throughout the world.
Combustion for power generation	20–>100 MWe	20–40% (electrical)	2.500–1600	Well established technology, especially deployed in Scandinavia and North America; various advanced concepts using fluid bed technology giving high efficiency, low costs and high flexibility. Commercially deployed waste to energy (incineration) has higher capital costs and lower (average) efficiency.
Co-combustion of biomass with coal	Typically 5–100 MWe at existing coal-fired stations. Higher for	30–40% (electrical)	100–1000 + costs of existing power station (depending on	Widely deployed in various countries, now mainly using direct combustion in combination with biomass

(Continued)

Table 9.3 Overview of current and projected performance data for the main conversion routes of biomass to power and heat and summary of technology status and deployment. Due to the variability of technological designs and conditions assumed, all costs are indicative.—cont'd

Conversion Option	Typical Capacity	Net Efficiency (LHV basis)	Investment Cost Ranges (€/kW)	Status and Deployment
	new multifuel power plants		biomass fuel + co-firing configuration)	fuels that are relatively clean. Biomass that is more contaminated and/or difficult to grind can be indirectly co-fired, e.g., using gasification processes. Interest in larger biomass co-firing shares and utilization of more advanced options is increasing.

(van Loo & Koppejan 2002; Knoef 2005; USDOE 1998; Dornburg & Faaij 2001).

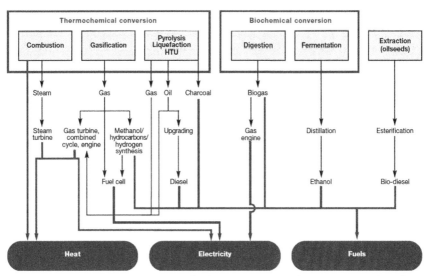

Figure 9.2 *Main conversion options for biomass to secondary energy carriers.* *(WEA 2000).*

thermochemical gasification can provide considerably more useful energy per hectare than the production of ethanol from starch or sugar crops and vegetable oils like RME. However, methanol and hydrogen derived from biomass are likely to be much more costly than conventional hydrocarbon fuels, unless oil prices rise to a level that is far higher than expected prices in coming decades. The most promising thermochemical conversion technology of lignocellulose raw materials currently available seems to be the production of pyrolytic oil or "bio-oil" (sometimes called bio-crude oil). This liquid can be produced by flash or fast pyrolysis processes at up to 80 percent weight yield. It has a heating value of about half that of conventional fossil fuels, but can be stored, transported and utilized in many circumstances where conventional liquid fuels are used, such as boilers, kilns and possibly turbines. It can be readily upgraded by hydrotreating or zeolite cracking into hydrocarbons for more demanding combustion applications such as gas turbines, and further refined into gasoline and diesel for use as transport fuels. In addition, there is considerable unexploited potential for the extraction and recovery of specialized chemicals. The technologies for producing bio-oil are evolving rapidly with improving process perfor-mance, larger yields and better quality products. Catalytic upgrading and extraction also show considerable promise and potential for transport fuels and chemicals; this is at a much earlier stage of development with more fundamental research underway at several laboratories. The utilization of the products is of major importance for the industrial realization of these technologies, and this is being investigated by several laboratories and companies. The economic viability of these processes is very promising in the medium term, and their integration into conventional energy systems presents no major problems. Charcoal is manufactured by traditional slow pyrolysis processes, and is also produced as a by-product from flash pyrolysis. It can be used industrially as a solid specialty fuel/reductant, for liquid slurry fuels, or for the manufacture of activated charcoal.

Biofuels have an average energy content that is approximately equal to that of brown coal (20 MJ/kg). There is no expected environmental impact or any disadvantage created by mixing some percentage of alcohol with diesel or gasoline fuel. There are no legal constraints at present.

9.3. CHARCOAL

There are several methods for processing wood residues to make them cleaner and easier to use as well as easier to transport. Production of charcoal is the most common. It is worth mentioning at this point that the conversion of woodfuel to charcoal does not increase the energy content of the fuel. Charcoal is often produced in rural areas and transported for use in urban areas. The wood is heated in the absence of sufficient oxygen and full

Figure 9.3 *Early charcoal production in earth kilns. (US Dept. of Agriculture, Forest Service 1961).*

combustion does not occur. This allows pyrolysis to take place, driving off the volatile gases and leaving the carbon or charcoal. The removal of the moisture means that the charcoal has much higher specific energy content than wood. Other biomass residues such as millet stems or corncobs can also be converted to charcoal.

Charcoal is produced in a kiln or pit. A typical traditional earth kiln will comprise the fuel to be carbonized, which is stacked in a pile and covered with a layer of leaves and earth. Once the combustion process is underway

Figure 9.4 *Charcoal production in Sosa. (©Jörg Behmann 2005).*

Figure 9.5 *Charcoal kiln, Kenya. (©Heinz Muller/ITDG/Practical Action 2010).*

the kiln is sealed. The charcoal can be removed when the process is complete and cooling has taken place.

A simple improvement to the traditional kiln is adding a chimney and air ducts which allow for a sophisticated gas and heat circulation system. There is very little capital investment and a significant increase in yield is achieved (Biomass [Technical Brief], 2010).

9.4. BRIQUETTES

Briquetting is the compression of loose biomass material. Many waste products, such as wood residues and sawdust from the timber industry, municipal waste, bagasse from sugar cane processing, or charcoal dust are briquetted to increase compactness and transportability. Briquetting is often a large-scale commercial activity and often the raw material will be carbonized during the process to produce a usable gas and also a more user-friendly briquette.

9.5. PELLETS

Pellets are more highly compressed loose biomass material than briquettes (ca. 650 kg/m^3, diameter 6 mm, length 30–40 mm, ash contents 1% and water contents less than 10%). Pellets are an important and rapidly growing biofuel for the production of heat and electricity. Pellets are developing quickly

Figure 9.6 *Certified pellets of high quality.* (Tom Bruton, 2003). (See color plate 24.)

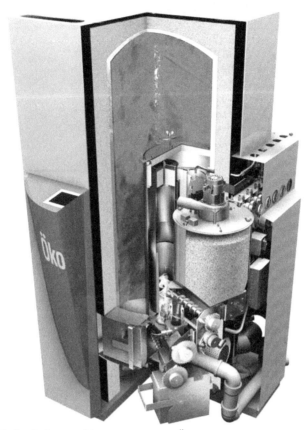

Figure 9.7 *Pellet boilers and heating systems. (ÖkoFEN, The Organic Energy Company 2012).*

in the coming years, to the benefit of the environment and local economies. Increased utilization of wood, agricultural residues, straw and industrial by-products in the Nordic countries, all over Europe and in North America indicates pellets have proven to be a realistic alternative to fossil fuels.

The substitution of fossil fuels with pellets, for heating of buildings and co-generation of electricity, can play an important role in fulfilling the Kyoto commitments for a secure supply of energy.

The potential for expansion of the pellet industry is significant. The Swedish Pellet Producers Association has estimated a market in Europe of 4–5 million tons per year within the next 5 years. Pellets will make an important contribution to the world strategy for renewable energy sources.

Figure 9.8 *Biogas injection in the gas pipeline.* (Biopact Team 2008).

9.6. BIOGAS

Biogas consists of similar proportions of methane and carbon dioxide and is produced by the anaerobic fermentation of wet organic feedstock in a process called *biomethanization*. Biomethanization has certain significance for the disposal of organic residues and waste products in the processing of agricultural products and in animal husbandry. Here, environmentally relevant aspects are of primary concern. The cultivation of plants with the goal of producing biogas is hardly practiced, though there are a number of green plants that are suitable in their fresh or in their ensiled form for biogas production. Through biomethanization, a broad spectrum of green plants could be used for energy because, in contrast to combustion, the raw material could be used in its natural moist state. This cannot be put into practice, however, because a number of fundamental conditions are not met. These will not be considered here. The plant species most suitable for the production of biogas are those that are rich in easily degradable carbohydrates, such as sugar and protein matter.

According to investigations by Zanuer and Küntzel (1986), the methane yields from maize, reed canary grass and perennial rye grass after silation and fermentation were identical.

Raw materials from lignocellulose-containing plant species are hardly suitable for the production of biogas. Biogas production from green plants has turned out to be very complicated, and the control of the fermentation process is very expensive and awkward. Besides this, for an acceptable yield of biogas from vegetable raw materials, production on a continuous, long term basis and a homogeneous substrate would be indispensable. The biogas yields from several plant species after 20 days retention time are given in Table 9.4.

Table 9.4 Biogas yield from fresh green plant materials, 20 days retention time

plant material	Biogas Yield (m³/1t VS)*
Alfalfa	440−630
Clove	430−450
Grass	520−640
Jerusalem artichoke	480−590
Maize plant	530−750
Sugarbeet leaves	490−510
Sweet sorghum	640−670

*Volatile Solids
(Weiland 1997).

9.7. ETHANOL

Ethanol is ethyl alcohol obtained by sugar fermentation (sugarcane), or by starch hydrolysis or cellulose degradation followed by sugar fermentation, and then by subsequent distillation.

Obtaining alcohol from vegetable raw materials has a long tradition in agriculture. The fermentation of sugar derived from agricultural crops using yeast to give alcohol, followed by distillation, is a well-established commercial technology. Alcohol can also be produced efficiently from starch crops (wheat, maize, potato, cassava, etc.). The glucose produced by the hydrolysis of starch can also be fermented to alcohol. The goal of directed use of cellulose-containing biomass from agriculturally utilized species for producing alcohol has not yet been practiced on a large-scale engineering level. Production has been confined to the use of wood, residues and waste materials. Table 9.5 lists the ethanol production of several sugar or starch producing plants. Technologies that can produce fuel ethanol from lignocellulose are already available (Ingram et al. 1995).

The challenge today is to assemble these technologies into a commercial demonstration plant. Many wood wastes and residues are available as feed mat, while high-yielding woody crops will also provide lignocellulose at lower prices than agriculture crops. Approaches for the improvement of acetonebutanol-ethanol (ABE) fermentation have focused on the development of hyper-amylolytic and hypercellulolytic clostridial strains with an improved potential for biomass-to-butanol conversion. The developed strains produce approximately 60 percent more butanol than the parental wild type strain (Blaschek 1995). These findings suggest that the developed strains are stable and offer a significant economic advantage for use in commercial

Table 9.5 Different crops in their production of ethanol per tonne or per acre

Estimated Ethanol Yield per Tonne of Cereal Grain (Keep in Mind There Is Great Variability between Varieties within Each Type of Grain):

Winter Wheat:	392 liters/tonne
CPS & SWS Wheat:	382 liters/tonne
Triticale	382 liters/tonne
Durum	377 liters/tonne
CWRS & Rye:	364 liters/tonne
Corn — ethanol	400 l/tonne
Barley (hulless)	380 l/tonne
Oats (hulless)	353 l/tonne
Oats (regular)	317 l/tonne

From Survey of Literature an Estimate of Ethanol Yield per Acre from Various Crops Is as Follows (Depending on Fact that there Can Be Huge Differences in Yields from One Region, or Even Field, to the Next):

Sugar Beets	2700 l/ac
Sugar Cane	2500 l/ac
Corn	1500 l/ac
Winter wheat	800 l/ac
Wheat, Barley	600 l/ac
One bushel wheat = 10 liters ethanol	

(This information was mostly compiled by Solulski and Tarasoff in 1997 and was gleaned off the Alberta Ag website (Day 2008)).

fermentation processes for producing butanol. Bioethanol can be used as a fuel in different ways: directly in special or modified engines as in Brazil or in Sweden, blended with gasoline at various percentages, or after transformation into ethyl tertio butyl ether (ETBE). In the first case it does not need to be dehydrated, whereas blends and ETBE production need anhydrous ethanol. Moreover, ETBE production requires neutral alcohol with a very low level of impurities. ETBE is ether resulting from a catalytic reaction between bioethanol and isobutylene, just as MTBE is processed from methanol. MTBE is the most commonly known fuel ether, used in European gasoline since 1973 and still generally used throughout the world to improve gasoline. Even the aviation industry has become involved in using ethanol as fuel.

Bioethanol can also be the alcohol used for the production of biodiesel, thus giving ethyl esters instead of methyl esters; the other constituent is vegetable oil from rape, sunflower, soybeans or other sources. European regulations enable direct blending of bioethanol up to 5 percent and of ETBE up to 15 percent with other oxygenated products, according to a Directive dated December 5, 1985 concerning crude oil savings by utilization of substitute fuels. On the other hand, an ambitious and realistic vision for 2030, by the European Energy Commission, is that up to one-fourth of the EU's transport fuel needs could be met by clean and CO_2-efficient biofuels. In fact, oil companies and car manufacturers in Europe do not like bioethanol blends, primarily because of their lack of water tolerance and their volatility and the need to put a label on the pumps to inform the consumer.

However, oil companies and car manufacturers are interested in ETBE, which is very similar and slightly better than the MTBE they are used to. ETBE has a lower Reid vapor pressure (RVP) than MTBE, and a higher

Figure 9.9 *Ethanol plant using corn, 2004.* *(DOE's Office of Energy Efficiency and Renewable Energy (EERE) 2008).*

octane value and lower oxygen content. Moreover, most MTBE producers can readily convert their plants to make ETBE after some process and equipment modifications.

ETBE and MTBE blended gasoline have several benefits:

- the oxygen content reduces carbon monoxide emissions
- the lower RVP lessens the pollution that forms ozone
- the high octane value reduces the need for hydrocarbon-based aromatic octane enhancers like benzene, which is carcinogenic

This growing interest in one product but not the other brought bioethanol producers into discussions with MTBE producers, and a partnership was initiated. There were only two of them in France: ELF and ARCO. ELF made a first test of ETBE production over five days in 1990 at its refinery in Feyzin near the town of Lyon, and a second test during a one-month period in 1992. The result was conclusive, so the company started ETBE production on an industrial scale in 1995. Oxygenated fuel feedstock (ETBE and MTBE) are considered as convenient options for reaching the octane value requirements of gasoline fuels. Oxygenated compounds offer considerable benefits if they are mixed with gasoline: reduction in the amount of carbon monoxide emissions and unburnt hydrocarbons. There is a growing demand for oxygenates in Europe, as well as in the USA as a result of the Clean Air Act Amendments of 1990. Bioethanol can be used in an undiluted form or mixed with motor gasoline. *Gasohol* is the term used in the USA to describe a maize-based mixture of gasoline (90 percent) and ethyl alcohol (10 percent); it should not be confused with *gasoil*, an oil

Figure 9.10 *Panoramic view of the Costa Pinto production plant set up to produce both sugar and ethanol fuel and other types of alcohol. Piracicaba, Sao Paulo, Brazil.* (Mariodo 2008).

Figure 9.11 *A car powered by ethanol 1931 in Brazil. (Joaquim Cohen 2007).*

product used to fuel diesel engines. Although engines need to be adapted to use bioethanol at 100 percent, unadapted engines can run the mixed fuels. Bioethanol can be used to substitute for MTBE and is added to unleaded fuel to increase octane ratings. In Europe, the preferred percentage—as recommended by the Association of European Automotive Manufacturers (AEAM)—is a 5 percent ethanol or 15 percent ETBE mix with gasoline.

9.8. BIO-OILS

Vegetable oils and fats, in contrast to the simple carbohydrate building blocks glucose and fructose, exhibit a number of modifications to the molecule's structure and are also more valuable regarding energy. From the point of view of plant breeding and plant cultivation of energy crops, it is high oil yields per hectare rather than quality aspects that are of primary importance.

The use of vegetable oils as fuels is not new. Since ancient times, oils have also been used as a material for burning and for lighting (the oil lamp!). The inventor Rudolph Diesel used peanut oil to fuel one of his engines at the Paris Exhibition of 1900, and wrote in 1912 that: " "The use of vegetable oils for engine fuels may seem insignificant today, but such oils may become, in the course of time, as important as petroleum and the coal tar products of the present time" (Nitske & Wilson, 1965). Now, biodiesel is increasingly used as a transportation fuel (Figure 9.13).

Worldwide, there are more than 280 plant species with more or less considerable oil contents in their seeds, fruits, tubers and/or roots. The oil

Table 9.6 Fuel properties of vegetable oils and methylesters in comparison with diesel fuel

Fuel Property	Diesel Fuel	Sunflower Oil	Sunflower Methyleste	Rape Oil	Rape Methyleste	Linseed Oil	Linseed Methylester
Specific gravity (kg/dm^3)	0.835	0.924	0.88	0.916	0.88	0.932	0.896
Viscosity (cSt):							
at 20 °C	5.1	65.8		77.8	7.5	50.5	8.4
at 50 °C	2.6	34.9	4.22	25.7	3.8		
Heat of combustion:							
Gross (MJ/liter)	38.4	36.5	35.3	37.2	33.1	36.9	35.6
Net (MJ/liter)	35.4	34.1	33.0	34.3			
Cetane number	>45	33	45–51	44–51	52–56		
Carbon residue (%)	0.15	0.42	0.05	0.25	0.02		0.22
Sulphur (%)	0.29	0.01	0.01	0.002	0.002	0.35	0.24

(Ortiz-Cañavate 1994).

Figure 9.12 *Neste Oil biodiesel plant in Singapore.* *(Neste Oil Oyj via Thomson Reuters ONE).*

Figure 9.13 *Virgin train powered by biodiesel.* *(Chris McKenna 2006).*

can be extracted by compressing the seeds (sunflower, rapeseed, etc.) mechanically with screw presses and expellers, with or without preheating. Pressing with preheating allows for the removal of up to 95 percent of the oil, whereas without preheating the amount is considerably less. Extraction by the use of solvents is more efficient: It removes nearly 99 percent of the oil from the seeds, but it has greater energy consumption than the expeller. Table 9.7 looks at the seed and oil yields of various crops.

Raw vegetable oils must be refined prior to their utilization in engines or transesterification. The most widespread idea is that fuel must be adapted to

Table 9.7 Typical oil extraction from 100kg of oil seeds

Crop	Oil/100 kg Seeds
Castor Seed	50 kg
Copra	62 kg
Cotton Seed	13 kg
Groundnut Kernel	42 kg
Mustard	35 kg
Palm Kernel	36 kg
Palm Fruit	20 kg
Rapeseed	37 kg
Sesame	50 kg
Soybean	14 kg
Sunflower	32 kg

(Addison, K 2004 Journeytoforever.org).

present-day diesel engines and not vice versa. In order to meet the requirements of diesel engines, vegetable oils can be modified into vegetable oil esters (transesterification). The transesterification procedure includes the production of methylesters—RME (rape methylester) or SME (sunflower methylester)—and glycerol (glycerin) by the processing of plant oils (triglycerides), alcohol (methanol) and catalysts (aqueous sodium hydroxide or potassium hydroxide). Cracking is another option for modifying the triglyceride molecule of the oil (Pernkopf 1984). However, the cracking products are very irregular and more suitable for gasoline substitution; the process has to be conducted on a large scale, costs are considerable; and conversion losses are also significant. These negative aspects together with the much lower efficiency of gasoline engines make the cracking process of minor interest.

The Veba process is yet another procedure for converting the triglyceride molecule. During the refining of mineral oil to form the different conventional fuels (gasoline, diesel, propane, butane, etc.), up to 20 percent rapeseed oil is added to the vacuum distillate. The molecules are cracked and the mixture is treated with hydrogen.

The generated fuel molecules are not different from conventional fuel molecules. Advantages of the Veba process are that no glycerin is produced as a byproduct, and that the generated fuel is not different from standardized fuels so there is no need for a special distribution system and handling. But there are also disadvantages: the high consumption of valuable hydrogen, and reductions in biodegradability (Vellguth 1991). Pure plant oils, particularly when refined and deslimed, can be used in pre-chamber (indirect

Figure 9.14 *The first flight by a commercial airline to be powered partly by biofuel has taken place. A Virgin Atlantic jumbo jet has flown between London's Heathrow and Amsterdam using fuel derived from a mixture of Brazilian babassu nuts and coconuts. (BBC News 2008).*

injected, like the Deutz engine described below) and swirl-chamber (Elsbett) diesel engines as pure plant oil or in a mixture with diesel (Schrottmaier et al. 1988). Pure plant oil cannot be used in direct injection diesel engines, which are used in standard tractors and cars, because engine coking occurs after some hours of operation. The addition of a small proportion of plant oils to diesel fuel is possible for all engine types, but will also lead to increased deposits in the engines in the long term (Ruiz-Altisent 1994).

The Elsbett engine is a recently developed diesel engine variant with a *duothermic combustion system* that uses a special turbulence swirl chamber and runs on pure vegetable oils. Another engine, developed by Deutz, uses the principle of turbulence with indirect injection and can run on purified plant oils. Its consumption is about 6 percent higher than that of other diesel engines, but it has proved to be robust and reliable. Other sources (Lohner 1963; van Basshuysen 1989) indicate that the consumption level of indirect injection swirl chamber diesel engines is 10–20 percent higher in relation to comparable diesel engines with direct injection. Another engine, still in development, is the John Deere *Wankelmotor*, which works with multi-type fuels, including vegetable oils. The transesterification of vegetable oils permits the wide utilization of these oils in existing engines, either as a 100 percent substitute or in blends with mineral diesel oil. Although vegetable oil esters have good potential, are well suited for mixture with or replacement of diesel fuels, and are effective in eliminating injection problems in direct injection diesel engines, their use still creates problems.

9.9. CONVERSION SYSTEMS TO HEAT, POWER AND ELECTRICITY

Production of heat and electricity dominate current bioenergy use. At present, the main growth markets for bioenergy are the European Union, North America, Central and Eastern Europe, and Southeast Asia (Thailand, Malaysia, Indonesia), especially with respect to efficient power generation from biomass wastes and residues and for biofuels. Two key industrial sectors for application of state-of-the-art biomass combustion (and potentially gasification) technology for power generation are the paper and pulp sector and cane-based sugar industry. Power generation from biomass by advanced combustion technology and co-firing schemes is a growth market worldwide. Mature, efficient, and reliable technology is available to turn biomass into power. In various markets the average scale of biomass combustion schemes rapidly increases due to improved availability of biomass resources and economies of scale of conversion technology. Competitive performance compared to fossil fuels is possible where lower cost residues are available particularly in co-firing schemes, where investment costs can be minimal. Specific (national) policies such as carbon taxes or renewable energy support can accelerate this development.

Gasification technology (integrated with gas turbines/combined cycles) offers even better perspectives for power generation from biomass in the medium term and can make power generation from energy crops competitive in many areas of the world once this technology has been proven on a commercial scale. Gasification, in particular larger scale circulating fluidized bed (CFB) concepts, also offers excellent possibilities for co-firing schemes (EREC 2007).

9.10. COMBINED HEAT AND POWER (CHP)

9.10.1. Heat

Biomass is the renewable heat source for small, medium and large-scale solutions. Pellets, chips and various byproducts from agriculture and forestry deliver the feedstock for bioheat. Pellets in particular offer possibilities for high energy density and standard fuels to be used in automatic systems, offering convenience for the final users. The construction of new plants to produce pellets, the installation of millions of burners/boilers/stoves and appropriate logistical solutions to serve the consumers should result in a significant growth of the pellet markets. Stoves and boilers operated with chips, wood pellets and wood logs have been optimized in recent years with respect to efficiency and emissions. However, more can be achieved in this area. In particular, further improvements regarding fuel handling, automatic control and maintenance requirements are necessary.

Rural areas present a significant market development potential for the application of those systems. There is a growing interest in the district heating plants which currently are run mainly by energy companies and sometimes by farmers' cooperatives for small-scale systems. The systems applied so far generally use forestry and wood-processing residues but the application of the agro-residues will be an important issue in the coming years (EREC 2007).

Direct combustion of biomass is the established technology for converting biomass to heat at commercial scales. Hot combustion gases are produced when the solid biomass is burned under controlled conditions. The gases are often used directly for product drying, but more commonly the hot gases are channeled through a heat exchanger to produce hot air, hot water or steam. The type of combustor most often implemented uses a grate to support a bed of fuel and to mix a controlled amount of combustion air. Usually, the grates move in such a way that the biomass can be added at one end and burned in a fuel bed which moves progressively down the grate to an ash removal system at the other end. More sophisticated designs permit the overall combustion process to be divided into its three main stages—drying, ignition and combustion of volatile constituents, and burnout of char—with separate control of conditions for each activity being possible. A fixed grate may be used for low-ash fuels, in which case the fuel charge is input by a spreader-stoker that maintains an even bed and fuel distribution and hence optimum combustion conditions. Grates are well proven and reliable, and can tolerate a varied range in fuel quality (moisture content and particle size). They have also been shown to be controllable and efficient. The goal of reducing emissions is one of the driving forces behind current developments. This goal has also driven the development of fluidized bed technology, which is the main alternative to grate-based systems.

Significant improvement in efficiencies can be achieved by installing systems that generate both useful power and heat (cogeneration plants have a typical overall annual efficiency of 80–90 percent). CHP is generally the most profitable choice for power production with biomass if heat, as hot water or as process steam, is needed. The increased efficiencies reduce both fuel input and overall greenhouse gas emissions compared to separate systems for power and heat, and also realize improved economics for power generation where expensive natural gas and other fuels are displaced. The technology for medium-scale CHP from 400kW to 4MW is now commercially available in the form of the Organic Ranking Cycle (ORC) systems or steam turbine systems. The first commercially available units for small scale CHP (1–10kW) are just arriving on the market, and a break-through for the gasification of biomass in the size between 100 and 500kW might occur in a few years (EREC 2007).

9.10.2. Electricity

The use of biomass for power generation has increased over recent years mainly due to the implementation of a favorable European and national political framework. In the EU-25, electricity generation from biomass (solid biomass, biogas and biodegradable fraction of municipal solid waste) grew by 19 percent in 2004 and 23 percent in 2005 and will have a continued yearly projected growth rate of 5.2 percent through 2020. However, most biomass power plants operating today are characterized by low boiler and thermal plant efficiencies and such plants are still costly to build. The main challenge therefore is to develop more efficient lower-cost systems. Advanced biomass-based systems for power generation require fuel upgrading, combustion and cycle improvement, and better flue-gas treatment. Future technologies have to provide superior environmental protection at lower cost by combining sophisticated biomass preparation, combustion, and conversion processes with post-combustion cleanup. Such systems include fluidized bed combustion, biomass-integrated gasification, and biomass externally fired gas turbines.

9.11. STEAM TECHNOLOGY

This process consists of creating steam and then using the steam to power an engine or turbine for generating electricity. Even though the production of steam by combustion of biomass is efficient, the conversion of steam to electricity is much less efficient. Where the production of electricity is to be maximized, the steam engine or turbine will exhaust into a vacuum condenser and conversion efficiencies are likely to be in the 5–10 percent range for plants of less than 1MWe, 10–20 percent for plants of 1 to 5MWe and 15–30 percent for plants of 5 to 25MWe. Low-temperature heat ($< 50°C$) is usually available from the condenser, though this is insufficient for most applications so it is normally wasted by dispersal into the atmosphere or a local waterway. The average conversion efficiency of steam plants in the USA is approximately 18 percent. The USA has installed 7000 MWe of wood fired plants since 1979.

Where there is a need for heat as well as electricity, for example the processing of biomass products such as the kilning of wood or the processing of sugar, palm oil, etc., the plant can be arranged to provide high-temperature steam. This is achieved by taking some steam directly from the boiler, by extracting partially expanded steam from a turbine designed for this purpose, or by arranging for the steam engine or turbine to produce exhaust steam at the required temperature.

All three options significantly reduce the amount of electricity available from the plant, though the overall energy efficiency may be much higher, 50–80 percent being common (Dumbleton 1997). Bioelectricity, particularly in the form of combined heat and power (CHP), is in the mainstream of current technological trends. Aside from the large number of small individual heating systems, more than 1000 MW of bioelectricity are currently produced annually by European utilities through conventional or advanced combustion (Palz 1995). Biomass, mainly in the form of industrial and agricultural residues and municipal solid waste, is presently used to generate electricity with conventional steam turbine power systems. The USA has an installed biomass (not just wood) electricity generating capacity of more than 8000 MW (Williams 1995). Although the power plants are small, typically 20 MW or less, and relatively capital intensive and energy inefficient, they can provide cost-competitive power where low-cost biomass is available, especially in combined heat and power applications. However, the use of this technology will not expand considerably in the future because low-cost biomass supplies are limited. Less capital-intensive and more energy-efficient technologies are needed to make the more abundant, but more costly, biomass sources usable. This involves the production of biomass grown on plantations dedicated to energy crops, which will be competitive particularly for power-only applications (Williams 1995).

Higher efficiency and lower unit capital costs can be realized with cycles involving gas turbines. Present developmental efforts are focused on biomass integrated gasifier/gas turbine (BIG/GT) cycles (Williams & Larson 1993).

In all of the cases so far described, steam technology can be considered as sound and well proven, though expensive. Electricity and CHP systems using biomass and steam technology can be competitive with electricity produced from fossil fuels in places where biomass residues are available at low or no cost, but prices will not be competitive if the biomass fuel has to be purchased at market prices. In this case, biomasses for electricity systems need other reasons for their utilization and electricity price structures that recognize these conditions. Steam technology will continue to be an acceptable alternative for biomass to electricity plants where these price structures exist. However, the environment and other benefits of using biomass are not maximized and continuous price support is needed, which are unacceptable factors in many instances.

The cost-effectiveness of electricity generation from biomass can be improved if conversion efficiencies increase and capital costs decrease. Raising conversion efficiencies also helps to maximize environmental benefits and associated environmental tax credits by decreasing the general dependency on fossil fuels. Because steam technology is in essence fully developed, there is unfortunately limited scope for finding improvements.

New conversion technologies are therefore important: gasification and pyrolysis are two relatively new technologies that are close to commercialization (Dumbleton 1997).

9.12. GASIFICATION

Gasification is a thermochemical process in which carbonaceous feedstock are partially oxidized by heating at temperatures as high as $1200°C$ to produce a stable fuel gas (Dumbleton 1997). By an exothermic chemical reaction with oxygen, a proportion of the carbon in the biomass fuel is converted to gas. The normal producer gas chemical reactions occur when the remainder of the biomass fuel is subjected to high temperatures in an oxygen-depleted atmosphere. The resulting fuel gas is mainly carbon monoxide, hydrogen and methane, with small amounts of higher hydrocarbons such as ethane and ethylene. These combustible gases are unfortunately reduced in quantity by carbon dioxide and nitrogen if air is used as the source of oxygen. Because carbon dioxide and nitrogen have no heating value, the heating value of the final fuel gas mixture is low: 4–6 MJ/Nm^3. This is equivalent to only 10–40 percent of the value for natural gas, usually 32 MJ/Nm^3, for which the current engines and turbines were created. The low heating value also makes pipeline transportation of the gas inappropriate. Unwanted byproducts such as char particles (unconverted carbon), tars, oils and ash are also present in small amounts. These could damage the engines and turbines and therefore must be removed or processed into additional fuel gas.

If oxygen is incorporated instead of air, the heating value of the gas is improved to 10–15 MJ/Nm^3. This improvement would permit unmodified engines and turbines to be used to generate electricity. However, the cost of oxygen production and the potential dangers associated with its utilization mean that the use of oxygen-blown gasifiers is not a favored option (Dumbleton, 1997).

Biomass may also be gasified under pressure, though it is not known if this will prove to be more cost-effective. Pressurized gasifiers are more expensive but will be smaller than the present normal gasifiers.

The pressurized gasifier system would be more efficient overall because tars will be more completely converted, sensible heat will be retained, and the need for fuel gas compression will be eliminated. However, pressure seals and the need to purge the feed system with inert gas will make the fuel feed system for pressurized gasifiers much more complicated. Such factors increase the capital and operating costs of pressurized gasifier plants, though a demonstration plant at Vernamo in Sweden uses this technology so it will be possible to evaluate these factors in the near future.

Direct combustion of the hot fuel gas from the gasifier in a boiler or furnace may only be possible for heating applications. The condensation of the tars that are cracked and burned in the combustor is prevented by maintaining high temperatures. If the biomass fuel or gasification process leads to the production of dust (ash), it may well be possible to remove this simply, using a hot gas cyclone. Some combustion systems will be able to withstand a limited dust loading with no gas clean-up necessary. Biomass gasification for combustion is often found in the pulp and paper industry, where waste products are utilized as fuel. There are also a number of Bioneer gasifiers on district heating schemes in Finland, though it is not clear whether this approach is more cost-effective than combustion.

The quality of the gas must be improved before it can be used in combustion engines or turbines, and the gas may have to be cooled to intermediate, if not low, temperatures because of temperature limitations in the engine's or turbine's fuel control systems. Reducing the gas temperature will increase the volumetric heating value of the gas, but it will also increase the condensation of tars, making the gas even less suitable for engine and turbines. A gas cleaning system will be essential in this case, and may consist of cyclones, filters or wet scrubbers. Wet scrubbers are particularly effective: in a single operation, they reduce the gas temperature and also capture tars (which are water soluble) and inert dust in the form of ash and mineral contaminants. However, a contaminated liquid waste stream is produced. This casts a shadow on the idea of biomass fuels and the concept of sustainability being clean alternatives.

Table 9.8 Power range and reported efficiencies of different technologies

Technology	Power Range (MWe) Efficiency (%)	Typical Overall
Steam engine	0.025 — 2.0	16
Steam turbine (back pressure)	1—150	25
Steam turbine (extraction-condensation cycle)	5—800	35
Steam turbine (condensation cycle)	1—800	40
Gas engine	0.025—1.5	27
Gas turbine	1.0—200	35
Integrated gas and steam combined cycle	5—450	55
Stirling engine	0.0003—?	40
Fuel cell	0.005—?	70

(Grimm 1996).

Fuels with ashes that have low ash softening and melting temperatures are inappropriate because of the high temperatures used during the gasification process.

Numerous annual crops and their residues fall into this category (Dumbleton 1997).

Advanced gasification for the production of electricity through integrated power generation cycles has a significant short-term potential, and is being developed in many parts of the world including Europe and North and South America. Both atmospheric and pressurized air gasification technologies are being promoted that are close coupled to a gas turbine. There are still problems to be resolved in interfacing the gasifier and turbine and meeting the gas quality requirements of the turbine fuel gas, but both areas are receiving substantial support. The production of hydrogen, methanol, ammonia and transport fuels through advanced gasification and synthesis are longer term objectives that rely on larger capacity plants that can take advantage of economies of scale.

9.12.1. Biomass Stoves

El Bassam and Forstinger constructed in 2003 a multifunctional oven for cooking, baking, grilling, and space heating using the gasification technology. It is a closed system without indoor smoke pollution.

Figure 9.15 *The multifunctional oven (MFO) for cooking, baking, grilling, and space heating.* (El Bassam and Forstinger 2003, personal communication). (See color plate 25.)

9.13. PYROLYSIS

Pyrolysis is the thermal degradation of carbonaceous material in the absence of air or oxygen. Temperatures in the 350–800°C range are most often used. Gas, liquid and solid products (char/coke) are always produced in pyrolysis

(a)   **(b)**

Figure 9.16a, b *Students enjoying a barbecue meal using MFO.* (El Bassam 2012). (See color plate 26.)

reactions, but the amounts of each can be influenced by controlling the reaction temperatures and residence time. The output of the desired product can be maximized by careful control of the reaction conditions. Because of this, heat is commonly added to the reaction indirectly. Charcoals for barbecues and industrial processes are the most common products of present pyrolysis. An overall efficiency of 35 percent by weight can be achieved by maximizing the output of the solid product through the implementation of long reaction times (hours to days) and low temperatures (350°C). Heat for the process in traditional kilns can be produced by burning the gas and liquid byproducts.

Higher temperatures and shorter residence times are used to produce pyrolysis oils. Optimum conditions are approximately 500°C with reaction times of 0.5 to 2 seconds. The necessary short reaction time is achieved by rapid quenching of the fuel, which prevents further chemical reactions from taking place. Preventing additional chemical reactions also allows higher molecular weight molecules to survive. The fast heating and cooling requirements of the raw material creates process control problems, which are most often dealt with by fine milling the feedstock to less than a few millimeters in size.

Conversion efficiency to liquid can be up to 85 percent by weight. Some gas is produced, and is commonly utilized in the fuel heating process. Solid char is also produced. The char remains in the pyrolysis oil, and must be removed before the oil can be used in combustion engines or turbines; it too can be used in the fuel heating process. The liquid product is a highly oxygenated hydrocarbon with a high water content, which is chemically and physically unstable. This instability may create difficulties in all aspects having to do with the storage and use of this product.

Detrimental chemical reactions within the liquid could also limit the maximum practical fuel storage time. This form of low temperature pyrolysis (450–550°C) is characterized by a reaction time of under one second. The main products produced are gas, bio-oil and coke. Figure 9.17 demonstrates the mass balance for this technology with respect to the initial weight of the biomass, using Miscanthus as an example.

The production of pyrolysis oil from biomass fuels has a number of potential advantages over gasification and combustion. The pyrolysis process can be separated from the final energy conversion process by both space and time.

9.14. METHANOL

Biomethanol is a fuelwood-based methyl alcohol obtained by the destructive distillation of wood or by gas production through gasification.

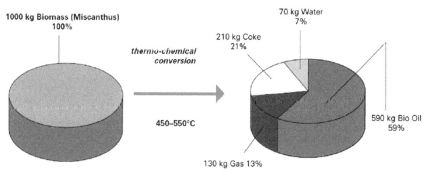

Figure 9.17 *Mass balance of miscanthus biomass conversion using flash or fast pyrolysis. (Shakir 1996).*

Biomethanol is a substitute product for synthetic methanol and is extensively used in the chemical industry (methanol is generally manufactured from natural gas and, to a lesser extent, from coal).

9.15. SYNTHETIC OIL

Bio-oil is directly produced from biomass by fast pyrolysis. The oil has some particular characteristics but has been successfully combusted for both heat and electricity generation. The largest plant built today produces 2 tons per hour, but plants for 4 and 6t/h (equivalent to 6–10 MWe) are at an advanced stage of planning. In the long term, pyrolysis oil could be used as a transportation fuel.

9.16. FUEL CELLS

The different fuel cell types can be divided into two groups: low temperature, or first generation (alkaline fuel cell, solid polymer fuel cell and phosphoric acid fuel cell); and high temperature, or second generation (molten carbonate fuel cell and solid oxide fuel cell).

Low-temperature cells have been commercially demonstrated, but are restricted in their fuel supply and are not readily integrated into combined heat and power applications. The high-temperature cells can use a wide variety of fuels through internal reforming techniques, have a high efficiency, and can be integrated into combined heat and power systems.

Although fuel cell capital costs are falling, increasing environmental restrictions are required before fuel cells will begin to replace conventional technologies in either the power generation or the transport sectors (Williams & Campbell 1994).

Methanol (MeOH) and hydrogen (H2) derived from biomass have the potential to make major contributions to transport fuel requirements by competitively addressing all of these challenges, especially when used in fuel cell vehicles (FCVs). In a fuel cell, the chemical energy of fuel is converted directly into electricity without first burning the fuel to generate heat to run a heat engine. The fuel cell offers a quantum leap in energy efficiency and the virtual elimination of air pollution without the use of emission control devices. Dramatic technological advances, particularly for the proton exchange membrane (PEM) fuel cell, focused attention on this technology for motor vehicles. The FCV has the potential to compete with the petroleum fueled ICEV (internal combustion engine vehicle) on both cost and performance grounds (AGTD 1994; Williams 1993, 1994), while effectively addressing air quality, energy security and global warming concerns.

One of the newest markets being looked at for bioethanol uses is fuel cells. Electrochemical fuel cells convert the chemical energy of bioethanol directly into electrical energy to provide a clean and highly efficient energy source. Bioethanol is one of the most ideal fuels for a fuel cell, besides the fact that it comes from and forms a catalyst deactivation within the fuel cell, which limits its life expectancy. Extensive research activities ensure that bioethanol remains among the most desirable fuels for fuel cells, delivering all the benefits that the bioethanol fuel cell technologies promise. Ballard Power Systems, Inc., of Vancouver, Canada, introduced a prototype PEM fuel cell bus in 1993 and planned to introduce PEM fuel cell buses on a commercial basis beginning in 1998. In April 1994, Germany's Daimler Benz introduced a prototype PEM fuel cell light duty vehicle (a van) and announced plans to develop the technology for commercial automotive applications. In the USA, the FCV is a leading candidate technology for accelerated development under the 'Partnership for a New Generation of Vehicles', a joint venture launched in September 1993 between the Clinton Administration and the US automobile industry. The partnership's goal is to develop in a decade production-ready prototypes of advanced, low polluting, safe cars that could be run on secure energy sources, especially renewable sources, that would have up to three times the fuel economy of today's gasoline ICEVs of comparable performance, and would cost no more to own and operate (Williams et al. 1995).

9.17. THE STIRLING ENGINE

The Stirling engine's name comes from its implementation of the Stirling cycle. Nitrogen or helium gas (Weber 1987) is shuttled back and forth between the hot and cold ends of the machine by the displacer piston. The power piston, with attached permanent magnets, oscillates within the linear alternator to generate electricity. Research in 1997 by Sunpower Inc. indicated that the Stirling engine's generator efficiently converted rough biomass into electricity with intrinsic load matching capacity.

A single engine produces 2.5 kW of 60 Hz, 120V alternating current power, and up to four engines may be used with a single burner. It is proposed for use at domestic and light industrial sites, and may be used for cogeneration, delivering both electricity and heat.

Positive features of the engine include the direct conversion of biomass heat into electricity, and an integrated alternator with greater than 90 percent efficiency. Further development of these types of engines is necessary and will open significant future opportunities to produce electricity directly from biomass.

9.18. ALGAE

Microalgae comprise a vast group of photosynthetic, heterotrophic organisms, which have an extraordinary potential for cultivation as energy crops. They can be cultivated under difficult agroclimatic conditions and are able to produce a wide range of commercially interesting by-products such as fats, oils, sugars and functional bioactive compounds. As a group, they are of particular interest in the development of future renewable energy scenarios. Certain microalgae are effective in the production of hydrogen and oxygen through the process of biophotolysis, while others naturally manufacture hydrocarbons suitable for direct use as high-energy liquid fuels. Microalgae can be grown both in open culture systems such as ponds, lakes and raceways, and in highly controlled closed culture systems similar to those used in commercial fermentation processes. Certain microalgae are very suitable for open system culture where the environmental conditions are very specific, such as high salt or high alkaline ponds, lakes or lagoons. The extreme nature of these environments severely limits the growth of competitive species, though other types of organisms may contaminate the culture. The advantages of such systems are that they generally require low investment, and are very cost effective and easy to manage. Closed-culture systems, on the other hand, require significantly higher investments and operating costs, but are independent of all variations in agroclimatic conditions and are very closely controlled for optimal performance and quality.

Algae grow rapidly, are rich in vegetable oil and can be cultivated in ponds of seawater, minimizing the use of fertile land and fresh water. Algae can double their mass several times a day and produce at least 15 times more oil per hectare than alternatives such as rape, palm soya or Jatropha. Facilities can be built on coastal land unsuitable for conventional agriculture. In the long term, algae cultivation facilities also have the potential to absorb waste carbon dioxide directly from industrial facilities such as power plants. Oil companies (Shell), DOE (USA) and other institutions are intensifying research activities in this field. Shell and a Hawaii-based algal biofuels company, HR Bio-petroleum, have formed a joint venture to grow marine algae for conversion into biodiesel, and some companies have far surpassed the 15,000 gallon per acre accepted benchmark. In fact, one company can produce 180,000 gallons of biodiesel every year from just one acre of algae. That comes to about 4000 barrels, at a cost of US $25 per barrel or US $0.59 per gallon.

9.18.1. Algae Bioreactors

Algal production systems have long been recognized as the most efficient means of producing biomass for food or fuel; they do not require arable land and therefore don't compete for space with existing crops. Over the same area, microalgae can produce 20–300 times more biodiesel than traditional crops (Table 9.9) and the remaining algal cake can still be useful for animal feed, fertilizer or other biofuel production systems.

However, the initial set-up and maintenance of such systems has, to date, always proven to be cost prohibitive for fuel.

Table 9.9 Comparison between crop efficiencies for biodiesel production

Plant Source	Biodiesel (l/ha/y)	Area Required to Match Current Global Oil Demand (Million ha)	Area Required as a Percentage of Global Land Mass
Soybean	446	10932	72.9
Rapeseed	119	4097	27.3
Mustard	1300	3750	25.0
Jatropha	1892	2577	17.2
Palm Oil	5950	819	5.5
Algae (low)	45000	108	0.7
Algae (high)	137000	36	0.2

(Handbook of Bioenergy Crops 2010).

Figure 9.18 *Conceptual rendering of a large-scale algae farm. http://www. algaeatwork.com/downloads/.*

In contrast to conventional crop plants, which yield a harvest once or twice a year, the microalgae have a life cycle of around 1–10 days depending on the process, with the result that multiple or continuous harvests with increased yield can be produced.

Figure 9.19 *Demonstration field photobioreactors: ASU-Arizona State University, Algal-based Biofuels & Biomaterial.* (Christine Lambrakis, Arizona State University (ASU)). (See color plate 27.)

(a)   **(b)**

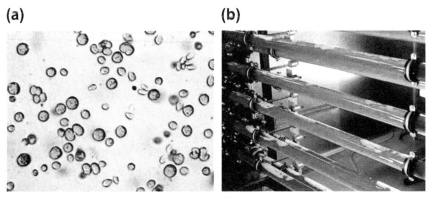

Figure 9.20 *(a) Chlamydomonas rheinhardtii green algal cells. (b) Tubular photo-bioreactor.* (Photographs courtesy of Ben Hankamer and Clemens Posten, Solar Biofuels Consortium: www.solarbiofuels.org). (See color plate 28.)

9.19. HYDROGEN

Hydrogen is viewed by many as the most promising fuel for light-duty vehicles (LDVs) for the future. Hydrogen can be produced through a large number of pathways and from many different feedstocks (both fossil and renewable). A key issue in evaluating the sustainability of a hydrogen fuel cell vehicle (FCV) is an analysis of the processes employed to produce the hydrogen and the efficiency of its use in the vehicle, the "well-to-wheel" (WTW) activities.

Several recent WTW studies, which include conventional as well as alternative fuel/propulsion system LDVs, are examined and compared. One potentially attractive renewable feedstock for hydrogen is biomass. A biomass to hydrogen pathway and its use in a FCV has only recently been included in WTW studies. The analysis is based on those studies which include biomass-derived hydrogen, comparing it to gasoline/diesel internal combustion engine vehicles (ICEVs) and FCVs which utilize hydrogen from other feedstock (natural gas and wind-generated electricity). Since hydrogen is not commercially produced from biomass, all of these studies utilize process and emissions data based on research results and extrapolations to commercial scale. We find that direct comparison of results between studies is challenging due to the differences in study methodologies and assumptions concerning feedstock, production processes, and vehicles. Overall, however, WTW results indicate that hydrogen produced via the gasification of biomass and its use in an FCV has the potential to reduce greenhouse gas

(GHG) emissions by between 75 and 100 percent and utilize little fossil energy compared to conventional gasoline ICEVs and hydrogen from natural gas FCVs (Fleming et al. 2005).

Shell Hydrogen LLC and Virent Energy Systems, Inc. announced in 2007 a five-year joint agreement to further develop and commercialize Virent's BioForming technology platform for hydrogen production. Virent's technology enables the economic production of hydrogen, among other fuels and chemicals, from renewable glycerol and sugar-based feedstock. The vast majority of hydrogen today is produced using fossil fuels, including natural gas and coal (renewableenergyworld.com, 2007). Scientists, in a published report in the new journal *ChemSusChem*, claim they have created an entirely natural and renewable method for producing hydrogen to generate electricity which could drastically reduce the dependency on fossil fuels in the future (El Bassam, N 2010). The breakthrough means ethanol, which comes from the fermentation of crops, can be completely converted to hydrogen and carbon dioxide for the first time. They created the first stable catalyst which can generate hydrogen using ethanol produced from crop fermentation at realistic conditions. The catalyst is made of very small nanoparticles of metals deposited on larger nanoparticles of a support called cerium oxide which is also used in catalytic converters in cars. At present the generation of hydrogen needed to power a mid-size fuel cell can be achieved using 1kg of this catalyst. As with traditional methods of hydrogen production, carbon dioxide is still created during the process they have developed. However, unlike fossil fuels which are underground, they are using ethanol generated from an above-the-ground source—plants or crops. This means that any carbon dioxide created during the process is assimilated back into the environment and is then used by plants as part of their natural cycle of growth (rdmag.com).

REFERENCES

BBC News, 2008. Airline in first biofuel flight [WWW]. BBC News. Available from: http://news.bbc.co.uk/2/hi/7261214.stm.

Biopact Team, 2008. Report: Biogas can replace all EU natural gas imports [WWW]. Biopact. Available from: http://news.mongabay.com/bioenergy/2008/01/report-biogas-can-replace-all-eu.html.

Department of Energy (DOE) Office of Energy Efficiency and Renewable Energy (EERE), https://www.eere-pmc.energy.gov/PMC_News/EERE_Program_News_3-08.aspx.

El Bassam, N. (Ed.), 1998. Energy plant species: Their use and impact on environment and development. Routledge, Taylor & Francis, Inc., Oxford.

El Bassam, N., 2010. Handbook of Bioenergy Crops: A Complete Reference to Species. Development and Applications Routledge, Taylor & Francis Group Ltd, Oxford.

Howtopedia, 2010. Biomass (Technical Brief) [Online]. Available from http://en: howtopedia.org/wiki/Biomass_%28Technical_Brief%29.

FURTHER READING

AGTD, (Allison Gas Turbine Division) Off General Motors Corporation, 1994. Research and development of proton-exchange membrane (PEM) fuel cell system for transportation applications. Initial Conceptual Design Report prepared for the Chemical Energy Division of Argonne National Laboratory, US Dept. of Energy.

Berndes et al. (2003); Hoogwijk et al. (2005a); Smeets et al. (2007).

Blaschek, H.P. 1995. Recent developments in the ABE fermentation. In: Workshop on Energy from Biomass and Wastes. Dublin Castle, Ireland, p. 28.

Dumbleton, F. 1997. Biomass conversion technologies: An overview. Aspects of Applied Biology 49, 341–347. Biomass and energy crops.

Fleming, J.S., Habibi, S., MacLean, H.L., Brinkman, N. 2005. Evaluation sustainability of producing hydrogen from biomass through well-two-wheel analysis. SAE Transactions vol. 114 (no 5), 729–745.

EREC 2007 Renewable Energy Technology Roadmap 20% by 2020. www.erec. orgGrimm, (1996), Handbook of Bioenergy Crops 2010.

Ingram, L.O., Bothast, R.J., Doran, J.B., Beall, D.S., Brooks, T.A., Wood, B.E., Lai, X., Asghari, K., Yomano, L.P. 1995. Genetic engineering of bacteria for the conversion of lignocellulose to ethanol. In: Procedings, Workshop on Energy from Biomass and Wastes. Dublin Castle, Ireland, p. 17.

Lohner, K. (1963) Die Brennkraftmaschine, VDI Verlag, Düsseldorf; van Basshuysen, R. (1989). Audi Turbodieselmotor mit Direkteinspritzung. Motortechnische Zeitschrift 50(10): 458–65.

Nitske, W.R., Wilson, C.M. 1965. Rudolph Diesel, Pioneer of the Age of Power. University of Oklahoma Press. 122–3.

Ortiz-Cañavate, 1994. Handbook of Bioenergy Crops 2010.

Palz, W., 1995. Future options for biomass in Europe. In: Proceedings, Workshop on Energy from Biomass and Wastes. Dublin Castle, Ireland, p. 2.

Pernkopf, J. (1984) The commercial and practical aspects of utilizing vegetable oils as diesel fuel substitute. Bio-Energy 84 World Conference, 18–21 June, Gothenburg, Sweden.

Renewable Energy World.com 2007. Shell Invests in Hydrogen from Biomass Production, Houston, Texas, http://www.renewableenergyworld.com/rea/news/article/2007/08/shell-invests-in-hydrogen-from-biomass-production-49525

Ruiz-Altisent, M. (Ed.), 1994. Biofuels – Application of Biologically Derived Products As Fuels Or Additives in Combustion Engines, p. 185. Directorate-General XII, Science, Research and Development, EUR 15647 EN.

Schrottmaier, J., Pernkopf, J., Wörgetter, M., 1988. Plant oil as fuel. A preliminary evaluation. Proceedings, KL Colloqium, 44–48.

Shakir, 1996. Handbook of Bioenergy Crops 2010.

Solulski and Tarasoff in 1997 (complied by), Alberta Ag Website (Day, 2008), Handbook of Bioenergy Crops, Routledge, Taylor & Francis, Inc., 2010.

U.S. Dept. of Agriculture, Forest Service, 1961. Charcoal Production, Marketing and Use, #2213 Available from. www.fpl.fs.fed.us/documnts/fplr/fplr2213.pdf.

van Loo and Koppejan (2002); Knoef (2005); USDOE (1998); Dornburg and Faaij (2001).

Vellguth, G., 1991. Energetische Nutzung von Rapsölmethyester. Dokumentation Nachwachsende Rohstoff, 17–21. FAL.

Weber, R. (1987) Stirling-motor treibt Wärmepumpe an. VDI-Nachrichtung Nr. 19.

Weiland, P., 1997. Personal communication.

Williams, B.C., Campbell, P.E., 1994. Application of fuel cells in 'clean' energy systems. Transactions of the Institution of Chemical Engineers B. Process Safety and Environmental Protection Transactions 72 (B), 252–256.

Williams, R H., 1995. The prospects for renewable energy. Siemens Review, 1–16.

Williams, R.H., Larson, E.D., 1993. Advanced gasification-based biomass power generation. In: Johansson, T.B., Kelly, H., Reddy, A.K.N., Williams, R.H. (Eds.), Renewable Energy: Sources for Fuels and Electricity. Island Press, Washington D.C., pp. 729–785.

Williams, R.H., Larson, E.D., Katofsky, R.E., Chen, J., 1995. Methonal and Hydrogen from Biomass for Transporation, with Comparisons to Methonal and Hydrogen from Natural Gas and Coal. The Centre for Energy and Environmental Studies, Princeton University. 47.

World Energy Assessment. 2004. Overview: 2004 Update, UNDP (United Nations Development Programme) (1997) Energy after Rio: Prospects and Challenges, Executive Summary, 35.

Zanuer, E., Küntzel, U., 1986. Methane production from ensiled plant material. Biomass 10, 207–223.

CHAPTER TEN

Hydropower

10.1. HYDROELECTRICITY

Hydropower, hydraulic power, hydrokinetic power, hydroelectricity or water power is power that is derived from the force or energy of falling water, which is then harnessed for useful purposes. For thousands of years, hydropower has been used for irrigation and the operation of various mechanical devices, such as watermills, sawmills, textile mills, dock cranes, and domestic elevators. In the last century, the term began to be associated with the modern development of hydro-electric power. This energy can be transmitted considerable distance between where it is created to where it is consumed (Figure 10.1).

Hydroelectricity is the term referring to electricity generated by hydropower, the production of electrical power through the use of the gravitational force of falling or flowing water. It is the most widely used form of renewable energy, accounting for 16 percent of global electricity consumption, and 3,427 terawatt-hours of electricity production in 2010, which continues the same rapid rate of increase experienced between 2003 and 2009.

Hydropower is produced in 150 countries (Figure 10.2), with the Asia-Pacific region generating 32 percent of global hydropower in 2010. China is the largest hydroelectricity producer, with 721 terawatt-hours of production in 2010, representing around 17 percent of domestic electricity use. There are now three hydroelectricity plants larger than 10 GW: the Three Gorges Dam in China, Itaipu Dam in Brazil, and Guri Dam in Venezuela (Worldwatch Institute (January 2012). "Use and Capacity of Global Hydropower Increases").

10.2. MICROHYDROPOWER SYSTEMS

Hydropower systems use the energy in flowing water to produce electricity or mechanical energy. Although there are several ways to harness the moving water to produce energy, *run-of-the-river systems*, which do not require large storage reservoirs, are often used for microhydropower systems.

Hydroelectric Dam

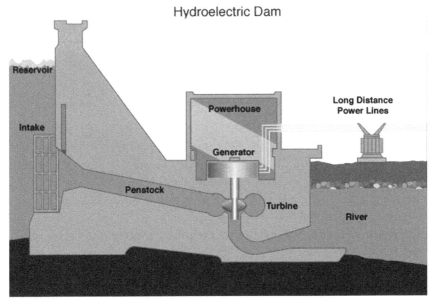

Figure 10.1 *A conventional dammed-hydro facility (Hydroelectric Dam) is the most common type of hydroelectric power generation. (Tennessee Valley Authority [TVA] 2005).*

Figure 10.2 Outflow during a test at the hydropower plant at the Hoover Dam, located on the Nevada-Arizona border. *(Photo courtesy U.S. Bureau of Reclamation).*

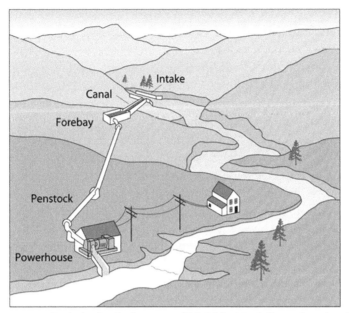

Figure 10.3 *Typical microhydro setup. (U.S. DOE 2011a).* (See color plate 29.)

For run-of-the-river microhydropower systems, a portion of a river's water is diverted to a water conveyance—channel, pipeline, or pressurized pipeline *(penstock)*—that delivers it to a turbine or waterwheel. The moving water rotates the wheel or turbine, which spins a shaft. The motion of the shaft can be used for mechanical processes, such as pumping water, or it can be used to power an alternator or generator to generate electricity (Figure 10.3).

A microhydropower system can be connected to an electric distribution system (grid-connected), or it can stand alone (off-grid) (U.S. DOE 2011b).

10.2.1. System Components

Run-of-the-river microhydropower systems consist of these basic components:

- Water conveyance—channel, pipeline, or pressurized pipeline (penstock) that delivers the water
- Turbine, pump, or waterwheel—transforms the energy of flowing water into rotational energy
- Alternator or generator—transforms the rotational energy into electricity
- Regulator—controls the generator
- Wiring—delivers the electricity

Commercially available turbines and generators are usually sold as a package. Do-it-yourself systems require careful matching of a generator with the turbine horsepower and speed.

Many systems also use an inverter to convert the low-voltage direct current (DC) electricity produced by the system into 120 or 240 volts of alternating current (AC) electricity. (Alternatively, you can buy household appliances that run on DC electricity.)

Whether a microhydropower system will be grid-connected or stand-alone will determine many of its *balance of system* components. For example, some stand-alone systems use batteries to store the electricity generated by the system. However, because hydropower resources tend to be more seasonal in nature than wind or solar resources, batteries may not always be practical for microhydropower systems. If you do use batteries, they should be located as close to the turbine as possible because it is difficult to transmit low-voltage power over long distances (U.S. DOE 2011a).

Small hydro is the development of hydroelectric power on a scale serving a small community or industrial plant. The definition of a small hydro project varies but a generating capacity of up to 10 megawatts (MW) is generally accepted as the upper limit of what can be termed small hydro. This may be stretched to 25 MW and 30 MW in Canada and the United States. Small-scale hydroelectricity production grew by 28% during 2008 from 2005, raising the total world small-hydro capacity to 85 GW. Over 70% of this was in China (65 GW), followed by Japan (3.5 GW), the United States (3 GW), and India (2 GW) (REN21 2006).

Small hydro plants may be connected to conventional electrical distribution networks as a source of low-cost renewable energy. Small hydro projects may also be built in isolated areas or in areas where there is no national electrical distribution network. With very small reservoirs and limited civil construction work, they have a relatively low environmental impact compared to a large hydro. This impact depends strongly on the balance between stream flow and power production (Figure 10.4).

Microhydro is a type of hydroelectric power that typically produces up to 100 kW of electricity using the natural flow of water. These installations can provide power to an isolated home or small community, or are sometimes connected to electric power networks. There are many of these installations around the world, particularly in developing nations as they can provide an economical source of energy without the purchase of fuel (Micro Hydro in the fight against poverty).

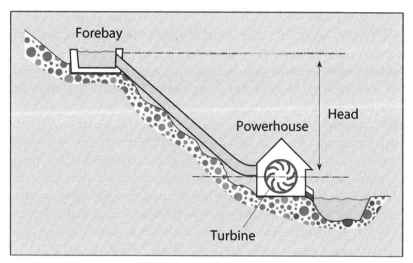

Figure 10.4 *Head is the vertical distance the water falls. Higher heads require less water to produce a given amount of power. (U.S. DOE 2011).* (See color plate 30.)

Microhydro systems complement photovoltaic solar energy systems in areas where water flow, and thus available hydro power, is highest in the winter when solar energy is at a minimum. Microhydro is frequently accomplished with an impulse turbine (pelton wheel) for a high head, low flow water supply. The installation is often a small dammed pool, at the top of a waterfall, with several hundred feet of pipe leading to a small generator housing.

Microhydro systems are very flexible and can be deployed in a number of different environments. They are dependent on how much water flow the source (creek, river, stream) has and the velocity of the flow of water. Energy can be stored in battery banks at sites that are far from a facility or used in addition to a system that is directly connected so that in times of high demand there is additional reserve energy available. These systems can be designed to minimize potential damage regularly caused by large dams or other mass hydroelectric generation sites (Research Institute for Sustainable Energy 2010).

10.3. TURBINE TYPES

"Several different types of water turbines can be used in microhydro installations, selection depending on the head of water, the volume of flow, and such factors as availability of local maintenance and transport of equipment to the site. For mountainous regions where a waterfall of 50 meters or more may be available, a Pelton wheel can be used. For low head installations, Francis or propeller-type turbines are used. Very low head installations of only a few meters may use propeller-type turbines in a pit. The very smallest microhydro installations may successfully use industrial centrifugal pumps, run in reverse as prime movers; while the efficiency may not be as high as a purpose-built runner, the relatively low cost makes the projects economically feasible.

In low-head installations, maintenance and mechanism costs often become important. A low-head system moves larger amounts of water, and is more likely to encounter surface debris. For this reason a Banki turbine, also called an Ossberger turbine, a pressurized self-cleaning cross flow waterwheel, is often preferred for low-head microhydropower systems. Though less efficient, its simpler structure is less expensive than other low-head turbines of the same capacity. Since the water flows in, then out of it, it cleans itself and is less prone to jam with debris (Wikipedia 2012c).

- Reverse Archimedes screw: two low-head schemes in England, Settle Hydro and Torrs Hydro, use an Archimedes screw, which is another debris-tolerant design. Efficiency 85%. (BusinessGreen, 2011)
- Gorlov: the Gorlov helical turbine free stream or constrained flow with or without a dam. (US DOE, 1998)
- Francis and propeller turbines. (Ashden Awards. "Micro-hydro". 2009)
- Kaplan turbine: an alternative to the traditional Kaplan turbine is a large diameter, slow turning, permanent magnet, sloped open flow VLH turbine with efficiencies of 90%. (http://www.vlh-turbine.com/EN/html/History.htm)
- Hydraulic water wheels and hydraulic wheel-part reaction turbine can have hydraulic efficiencies of 67% and 85% respectively.
- Gravitation water vortex power plant. Part of the river flow at a weir or natural water fall is diverted into a round basin with a central bottom exit that creates a vortex. A simple rotor (and connected generator) is moved by the kinetic energy. Efficiencies at 1/3 partial flow are calculated from 64% to 83%.

10.4. POTENTIAL FOR RURAL DEVELOPMENT

In relation to rural development, the simplicity and low relative cost of microhydro systems open up new opportunities for some isolated communities in need of electricity (Figure 10.5). With only a small stream needed, remote areas can access lighting and communications for homes,

Figure 10.5 *Small hydropower plant, 2011 (India Ministry of New and Renewable Energy 2011).* (See color plate 31.)

medical clinics, schools, and other facilities. Microhydro can even run a certain level of machinery supporting small businesses. Regions along the Andes Mountains and in Sri Lanka and China already have similar, active programs. One seemingly unexpected use of such systems in some areas is to keep young community members from moving into more urban regions in order to spur economic growth (The Ashden Awards for Sustainable Energy 2010). With the use of financial incentives for less carbon-intensive processes, the future of microhydro systems will become more appealing.

REFERENCES

Ashden Awards. Micro-hydro. Retrieved 2009-06-29.
Gorlov, A.M., 1998. Development of the helical reaction hydraulic turbine. Final Technical Report, The US DOE The Department of Energy's (DOE) Information Bridge: DOE Scientific and Technical Information.
Micro Hydro in the fight against poverty, http://tve.org/ho/doc.cfm?aid=1636&lang=English.
Renewables Global Status Report 2006 Update, REN21, published 2006.
Research Institute for Sustainable Energy, Microhydro, Retrieved 9 December 2010.
Shankleman, Jessica. The Queen's hydro energy scheme slots into place BusinessGreen, 2011. Retrieved 21 July 2012.
The Ashden Awards for Sustainable Energy, Micro-hydro, Retrieved 20 November 2010.
U.S. Department of Energy (DOE), 2011a. Microhydropower System Components [WWW]. Energy Savers. Available from: http://www.energysavers.gov/your_home/electricity/index.cfm/mytopic=11100.

U.S. Department of Energy (DOE), 2011b. How a Microhydropower System Works [WWW]. Energy Savers How a Microhydropower System Works, accessed. 2011. Available from http://www.energysavers.gov/your_home/electricity/index.cfm/mytopic=11060http://www.energysavers.gov/your_home/electricity/index.cfm/mytopic=11060.

Wikipedia 2012c. Microhydro systems [Online], Turbine Types, Available from http://en.wikipedia.org/wiki/Microhydro_systems.

CHAPTER ELEVEN

Marine Energy

Marine energy or marine power (also sometimes referred to as ocean energy or ocean power) refers to the energy carried by ocean waves, tides, salinity, and ocean temperature differences. The movement of water in the world's oceans creates a vast store of kinetic energy, or energy in motion. This energy can be harnessed to generate electricity to power homes, transport and industries.

The term *marine energy* encompasses both wave power, power from surface waves, and tidal power, obtained from the kinetic energy of large bodies of moving water (Ocean Energy Glossary 2007).

The oceans have a tremendous amount of energy and are close to many if not most concentrated populations. Ocean energy has the potential of providing a substantial amount of new renewable energy around the world (Carbon Trust 2006).

Marine current power is a form of marine energy obtained from harnessing of the kinetic energy of marine currents, such as the Gulfstream. Although not widely used at present, marine current power has an important potential for future electricity generation. Marine currents are more predictable than wind and solar power (How Stuff Works 2010).

Oceans cover more than 70% of the Earth's surface. As the world's largest solar collectors, oceans generate thermal energy from the sun. They also produce mechanical energy from the tides and waves. Even though the sun affects all ocean activity, the gravitational pull of the moon primarily drives the tides, and the wind powers the ocean waves.

11.1. OCEAN THERMAL ENERGY CONVERSION

A process called *ocean thermal energy conversion* (OTEC) uses the heat energy stored in the Earth's oceans to generate electricity.

OTEC works best when the temperature difference between the warmer, top layer of the ocean and the colder, deep ocean water is about 20°C (36°F). These conditions exist in tropical coastal areas, roughly between the Tropic of Capricorn and the Tropic of Cancer. To bring the cold water to the surface,

OTEC plants require an expensive, large diameter intake pipe, which is submerged a mile or more into the ocean's depths. Some energy experts believe that if it could become cost-competitive with conventional power technologies, OTEC could produce billions of watts of electrical power.

OTEC technology is not new. In 1881, Jacques Arsene d'Arsonval, a French physicist, proposed tapping the thermal energy of the ocean. But it was d'Arsonval's student, Georges Claude, who in 1930 actually built the first OTEC plant in Cuba. The system produced 22 kilowatts of electricity with a low-pressure turbine. In 1935, Claude constructed another plant aboard a 10,000-ton cargo vessel moored off the coast of Brazil. Weather and waves destroyed both plants before they became net power generators. (Net power is the amount of power generated after subtracting power needed to run the system.)

In 1956, French scientists designed another 3-megawatt OTEC plant for Abidjan, Ivory Coast, West Africa. The plant was never completed, however, because it was too expensive.

The United States became involved in OTEC research in 1974 with the establishment of the Natural Energy Laboratory of Hawaii Authority. The Laboratory has become one of the world's leading test facilities for OTEC technology.

11.2. TECHNOLOGIES

The types of OTEC systems are described in the following sections.

11.2.1. Closed-cycle

These systems use fluid with a low-boiling point, such as ammonia, to rotate a turbine to generate electricity. Warm surface seawater is pumped through a heat exchanger where the low-boiling-point fluid is vaporized. The expanding vapor turns the turbo-generator. Cold deep-seawater—pumped through a second heat exchanger—condenses the vapor back into a liquid, which is then recycled through the system.

In 1979, the Natural Energy Laboratory and several private-sector partners developed the mini OTEC experiment, which achieved the first successful at-sea production of net electrical power from closed-cycle OTEC. The mini OTEC vessel was moored 1.5 miles (2.4 km) off the Hawaiian coast and produced enough net electricity to illuminate the ship's light bulbs and run its computers and televisions.

In 1999, the Natural Energy Laboratory tested a 250-kW pilot OTEC closed-cycle plant, the largest such plant ever put into operation.

11.2.2. Open-cycle

These systems use the tropical oceans' warm surface water to make electricity. When warm seawater is placed in a low-pressure container, it boils. The expanding steam drives a low-pressure turbine attached to an electrical generator. The steam, which has left its salt behind in the low-pressure container, is almost pure fresh water. It is condensed back into a liquid by exposure to cold temperatures from deep-ocean water.

In 1984, the Solar Energy Research Institute (now the National Renewable Energy Laboratory) developed a vertical-spout evaporator to convert warm seawater into low-pressure steam for open-cycle plants; energy conversion efficiencies as high as 97% were achieved. In May 1993, an open-cycle OTEC plant at Keahole Point, Hawaii, produced 50,000 watts of electricity during a net power-producing experiment.

11.2.3. Hybrid

These systems combine the features of both the closed-cycle and open-cycle systems. In a hybrid system, warm seawater enters a vacuum chamber where it is flash-evaporated into steam, similar to the open-cycle evaporation process. The steam vaporizes a low-boiling-point fluid (in a closed-cycle loop) that drives a turbine to produce electricity.

11.2.4. Advantages and Benefits of OTEC Technology

OTEC has important benefits other than power production. For example, air conditioning can be a byproduct. Spent cold seawater from an OTEC plant can chill fresh water in a heat exchanger or flow directly into a cooling system. Simple systems of this type have air conditioned buildings at the Natural Energy Laboratory for several years.

OTEC technology also supports chilled-soil agriculture. When cold seawater flows through underground pipes, it chills the surrounding soil. The temperature difference between plant roots in the cool soil and plant leaves in the warm air allows many plants that evolved in temperate climates to be grown in the subtropics. The Natural Energy Laboratory maintains a demonstration garden near its OTEC plant with more than 100 different fruits and vegetables, many of which would not normally survive in Hawaii.

Aquaculture is perhaps the most well-known byproduct of OTEC. Cold-water delicacies, such as salmon and lobster, thrive in the nutrient-rich, deep seawater from the OTEC process. Microalgae such as Spirulina, a health food supplement, also can be cultivated in the deep-ocean water.

As mentioned earlier, another advantage of open or hybrid-cycle OTEC plants is the production of fresh water from seawater. Theoretically, an OTEC plant that generates 2-MW of net electricity could produce about 4,300 cubic meters (14,118.3 cubic feet) of desalinated water each day (U.S. Department of Energy EERE 2011).

OTEC also may one day provide a means to mine ocean water for 57 trace elements. Most economic analyses have suggested that mining the ocean for dissolved substances would be unprofitable. Mining involves pumping large volumes of water and the expense of separating the minerals from seawater. But with OTEC plants already pumping the water, the only remaining economic challenge is to reduce the cost of the extraction process (Ocean Thermal Energy Conversion 2011).

11.3. OCEAN TIDAL POWER

Some of the oldest ocean energy technologies use tidal power. All coastal areas consistently experience two high and two low tides over a period of slightly greater than 24 hours. For those tidal differences to be harnessed into electricity, the difference between high and low tides must be at least five meters, or more than 16 feet. There are only about 40 sites on the Earth with tidal ranges of this magnitude.

Currently, there are no tidal power plants in the United States. However, conditions are good for tidal power generation in both the Pacific Northwest and the Atlantic Northeast regions of the country.

Tidal power technologies include the following:
- Barrage or dam. A barrage or dam is typically used to convert tidal energy into electricity by forcing the water through turbines, activating a generator. Gates and turbines are installed along the dam. When the tides produce an adequate difference in the level of the water on opposite sides of the dam, the gates are opened. The water then flows through the turbines. The turbines turn an electric generator to produce electricity.
- Tidal fence. Tidal fences look like giant turnstiles. They can reach across channels between small islands or across straits between the mainland and an island. The turnstiles spin via tidal currents typical of

coastal waters. Some of these currents run at 5–8 knots (5.6–9 miles per hour) and generate as much energy as winds of much higher velocity. Because seawater has a much higher density than air, ocean currents carry significantly more energy than air currents (wind).

• Tidal turbine. Tidal turbines look like wind turbines. They are arrayed underwater in rows, as in some wind farms. The turbines function best where coastal currents run at between 3.6 and 4.9 knots (4 and 5.5 mph). In currents of that speed, a 15-meter (49.2-feet) diameter tidal turbine can generate as much energy as a 60-meter (197-feet) diameter wind turbine. Ideal locations for tidal turbine farms are close to shore in water depths of 20–30 meters (65.5–98.5 feet) (International Climate and Environmental Change Assessment Project [ICECAP] 2007).

Cork-based Ocean Energy Limited is now working with the Wave Hub team to secure the necessary consents and to build a full-scale wave energy device which could be deployed by the end of this year.

Ocean Energy's device, called an OE Buoy (Figure 11.1), uses the oscillating water column principle. As waves enter a sub-sea chamber, they force air through a turbine on the surface, generating electricity. The technology has only one moving part, minimizing maintenance costs.

Figure 11.1 *OE Buoy in Cornwall. (Ocean Energy from* ReNews Europe *2012).*

SeaGen was installed in the tidal currents of Strangford Lough, Northern Ireland in 2008 (Figure 11.2). However, there is a suite of rival designs racing to harness ocean energy.

Figure 11.2 *Largest tidal stream System installed.* *(*The Energy Blog, *2008).* (See color plate 32.)

The AK1000 tidal turbine, developed by Atlantis Resources, was installed at the European Marine Energy Centre in late 2010. Despite its large size, including an 18-meter rotor diameter, the unit has minimal effects on the surrounding ecosystem because of its slow rotational speed. They are also planning to install the turbine in Gujarat, India (Figure 11.3).

11.4. OCEAN WAVE POWER

Wave power devices extract energy directly from surface waves or from pressure fluctuations below the surface. Renewable energy analysts believe there is enough energy in the ocean waves to provide up to 2 terawatts of electricity. (A terawatt is equal to a trillion watts.)

Wave power can't be harnessed everywhere. Wave-power rich areas of the world include the western coasts of Scotland (Figure 11.4), northern

Figure 11.3 *Atlantis to install tidal power farm in Gujarat, India.* (Atlantis Resources Corp. at *Offshore Wind* 2012).

Canada, southern Africa, Australia, and the northeastern and northwestern coasts of the United States. In the Pacific Northwest alone, it's feasible that wave energy could produce 40–70 kilowatts (kW) per meter (3.3 feet) of western coastline. The West Coast of the United States is more than 1,000 miles long.

Wave energy can be converted into electricity through both offshore and onshore systems.

Figure 11.4 *The Oyster 1 wave energy converter, developed by Aquamarine Power, has been operating at the European Marine Energy Centre in Scotland since November 2009. The company plans to install a second unit at the testing center in 2011. (Lamb 2011).* (See color plate 33.)

11.4.1. Offshore Systems

Offshore systems are situated in deep water, typically of more than 40 meters (131 feet). Sophisticated mechanisms—like the Salter Duck—use the bobbing motion of the waves to power a pump that creates electricity. Other offshore devices use hoses connected to floats that ride the waves. The rise and fall of the float stretches and relaxes the hose, which pressurizes the water, which, in turn, rotates a turbine.

Specially built seagoing vessels can also capture the energy of offshore waves. These floating platforms create electricity by funneling waves through internal turbines and then back into the sea.

11.4.2. Onshore Systems

Built along shorelines, onshore wave power systems extract the energy in breaking waves. Onshore system technologies include the following:

*   Oscillating water column.
    The oscillating water column consists of a partially submerged concrete or steel structure that has an opening to the sea below the waterline. It encloses a column of air above a column of water. As waves enter the air column, they cause the water column to rise and fall. This alternately compresses and depressurizes the air column. As the wave retreats, the air is drawn back through the turbine as a result of the reduced air pressure on the ocean side of the turbine.
*   Tapchan.
    The tapchan, or tapered channel system, consists of a tapered channel, which feeds into a reservoir constructed on cliffs above sea level. The narrowing of the channel causes the waves to increase in height as they move toward the cliff face. The waves spill over the walls of the channel into the reservoir and the stored water is then fed through a turbine.
*   Pendulor device.
    The pendulor wave–power device consists of a rectangular box, which is open to the sea at one end. A flap is hinged over the opening and the action of the waves causes the flap to swing back and forth. The motion powers a hydraulic pump and a generator.

11.5. ENVIRONMENTAL AND ECONOMIC CHALLENGES

In general, careful site selection is the key to keeping the environmental impacts of wave power systems to a minimum. Wave energy system planners can choose sites that preserve scenic shorefronts. They also can avoid areas where wave energy systems can significantly alter flow patterns of sediment on the ocean floor.

Economically, wave power systems have a hard time competing with traditional power sources. However, the costs to produce wave energy are coming down. Some European experts predict that wave power devices will find lucrative niche markets. Once built, they have low operation and maintenance costs because the fuel they use—seawater—is free (U.S. Department of Energy [DOE] 2011a).

An Irish wave energy developer is to work with Cornwall's Wave Hub with a view to deploying the scheme's first device by the end of 2012.

Tidal power plants that dam estuaries can impede sea life migration, and silt build-ups behind such facilities can impact local ecosystems. Tidal fences may also disturb sea life migration. Newly developed tidal turbines may prove ultimately to be the least environmentally damaging of the tidal power technologies because they don't block migratory paths.

It doesn't cost much to operate tidal power plants, but their construction costs are high and lengthen payback periods. As a result, the cost per kilowatt-hour of tidal power is not competitive with conventional fossil fuel power.

OTEC power plants require substantial capital investment upfront. OTEC researchers believe private sector firms probably will be unwilling to make the enormous initial investment required to build large-scale plants until the price of fossil fuels increases dramatically or until national governments provide financial incentives. Another factor hindering the commercialization of OTEC is that there are only a few hundred land-based sites in the tropics where deep-ocean water is close enough to shore to make OTEC plants feasible (U.S. DOE 2011b).

REFERENCES

Carbon Trust, Future Marine Energy. Results of the Marine Energy Challenge: Cost competitiveness and growth of wave and tidal stream energy, January 2006. http://www.oceanrenewable.com/wp-content/uploads/2007/03/futuremarineenergy.pdf.
How Stuff Works, "Ocean currents". p. 4. Retrieved 2 November 2010. http://science.howstuffworks.com/environmental/earth/oceanography/ocean-current4.htm.
International Climate and Environmental Change Assessment Project (ICECAP), 2007. Tidal power: Clean, reliable and renewable accessed at: http://icecap.us/images/uploads/TIDAL_POWER.pdf.
Lamb, T., 2011. Driving ocean energy innovation in Scotland. Renewable Energy World Viewed June 30, 2012 at: http://www.renewableenergyworld.com/rea/news/article/2011/06/driving-ocean-energy-innovation-in-scotland.

Ocean Energy Glossary 2007. Wave Energy Centre supported by the Co-ordinated Action of Ocean Energy EU funded Project (CA-OE) and with the Implementing Agreement on Ocean Energy Systems (IEA-OES) (PDF)

Offshore Wind, 2012. Atlantis to install tidal power farm in Gujarat, India [WWW]. Available from: http://www.offshorewind.biz/2012/02/01/atlantis-to-install-tidal-power-farm-in-gujarat-india/.

ReNews Europe, 2012. OE wins Wave Hub deal [WWW]. Available from: http://renews.biz/story.php?page_id=71&news_id=1372.

The Energy Blog, 2008. Largest tidal stream system installed Weblog [Online] 06 April. Available from: http://thefraserdomain.typepad.com/energy/ocean_power/.

U.S. Department of Energy (DOE) EERE, 2011a. Ocean Wave Power accessed from: http://www.energysavers.gov/renewable_energy/ocean/index.cfm/mytopic=50009.

U.S. Department of Energy (DOE) EERE, 2011b. Ocean Thermal Energy Conversion accessed from: http://www.eere.energy.gov/basics/renewable_energy/ocean_thermal_energy_conv.html.

CHAPTER TWELVE

Geothermal Energy

12.1. ORIGIN OF GEOTHERMAL HEAT

Geothermal energy is thermal energy generated and stored in the Earth. Thermal energy is the energy that determines the temperature of matter. Earth's geothermal energy originates from the original formation of the planet (20%) and from radioactive decay of minerals (80%) (Turcotte 2002). The geothermal gradient, which is the difference in temperature between the core of the planet and its surface, drives a continuous conduction of thermal energy in the form of heat from the core to the surface (Figure 12.1). The heat that is used for geothermal energy can be stored deep within the Earth, all the way down to Earth's core—4,000 miles down.

At the core, temperatures may reach over 9,000 degrees Fahrenheit (5,000 degrees Celsius). Heat conducts from the core to surrounding rock. Extremely high temperature and pressure cause some rock to melt, which is commonly known as magma. Magma convects upward since it is lighter than the solid rock. This magma then heats rock and water in the crust, sometimes up to 700 degrees Fahrenheit (370 degrees Celsius) (Nemzer 2012).

From hot springs, geothermal energy has been used for bathing since Paleolithic times and for space heating since ancient Roman times, but it is now better known for electricity generation. Worldwide, about 10,715 megawatts (MW) of geothermal power is online in 24 countries. An additional 28 gigawatts (GW) of direct geothermal heating capacity is installed for district heating, space heating, spas, industrial processes, desalination and agricultural applications (Fridleifsson 2008).

Geothermal power is cost effective, reliable, sustainable, and environmentally friendly (Glassley 2010), but has historically been limited to areas near tectonic plate boundaries (Figure 12.2). Recent technological advances have expanded this range especially for applications such as home heating. Geothermal wells release greenhouse gases trapped deep within the Earth, but these emissions are much lower per energy unit than those of fossil fuels: Therefore, geothermal power has the potential to help mitigate global warming.

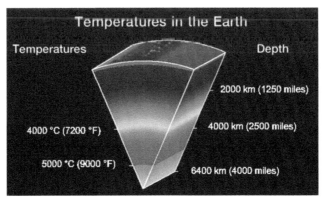

Figure 12.1 *Temperatures in the Earth.* *(Our Energy 2012).* (See color plate 34.)

The Earth's geothermal resources are theoretically more than adequate to supply humanity's energy needs, but only a very small fraction may be profitably exploited. Drilling and exploration for deep resources is very expensive. Forecasts for the future of geothermal power depend on assumptions about technology, energy prices, subsidies, and interest rates (Cothran 2002).

Figure 12.2 *Steam rising from the Nesjavellir Geothermal Power Station in Iceland.* *(Photo: Gretar Ívarsson 2007).* (See color plate 35.)

12.2. GEOTHERMAL ELECTRICITY

Heat from the Earth—geothermal energy—heats water that has seeped into underground reservoirs. These reservoirs can be tapped for a variety of uses, depending on the temperature of the water. The energy from high-temperature reservoirs (225°–600°F) can be used to produce electricity. Technologies in use include dry steam power plants, flash steam power plants and binary cycle power plants. Geothermal electricity generation is currently used in 24 countries (Geothermal Energy Association 2010), while geothermal heating is in use in 70 countries (Fridleifsson 2008).

Estimates of the electricity-generating potential of geothermal energy vary from 35 to 2,000 GW. Current worldwide installed capacity is 10,715 megawatts (MW) (Geothermal Energy Association 2010), with the largest capacity in the United States (3,086 MW), the Philippines, and Indonesia.

Geothermal power is considered to be sustainable because the heat extraction is small compared with the Earth's heat content (Rybach, Ladislaus 2007). The emission intensity of existing geothermal electric plants is on average 122 kg of CO_2 per megawatt-hour (MW·h) of electricity, about one-eighth of a conventional coal-fired plant (Bertani, Thain, 2002).

The International Geothermal Association (IGA) reported in 2010 that 10,715 megawatts (MW) of geothermal power in 24 countries were online, and which were expected to generate 67,246 GWh of electricity. This represented a 20% increase in online capacity since 2005. IGA projects growth to 18,500 MW by 2015, due to the projects presently under consideration, often in areas previously assumed to have little exploitable resource (Geothermal Energy Association, pages 4–6).

In 2010, the United States led the world in geothermal electricity production with 3,086 MW of installed capacity from 77 power plants (Geothermal Energy Association 2010). The largest group of geothermal power plants in the world is located at The Geysers, a geothermal field in California (Khan 2007). The Philippines is the second highest producer, with 1,904 MW of capacity online (Figure 12.3). Geothermal power makes up approximately 18% of the country's electricity generation (Geothermal Energy Association 2010).

Figure 12.3 *Palinpinon Geothermal power plant in Sitio Nasulo, Brgy. Puhagan, Valencia, Negros Oriental, Philippines. (Photo: Mike Gonzalez 2010).*

12.3. TYPES OF GEOTHERMAL POWER PLANTS

There are currently three types of geothermal power plants:

- **Dry steam plants** use steam from underground wells to rotate a turbine, which activates a generator to produce electricity. There are only two known underground resources of steam in the United States: The Geysers in northern California and Old Faithful in Yellowstone National Park. Since Yellowstone is protected from development, the power plants at The Geysers are the only dry steam plants in the country.
- **Flash steam plants**, the most common type of geothermal power plant, use water at temperatures greater than 360°F. As this hot water flows up through wells in the ground, the decrease in pressure causes some of the water to boil into steam. The steam is then used to power a generator and any leftover water and condensed steam is returned to the reservoir (Figure 12.4).
- **Binary cycle plants** use the heat from lower-temperature reservoirs (225°–360°F) to boil a working fluid, which is then vaporized in a heat

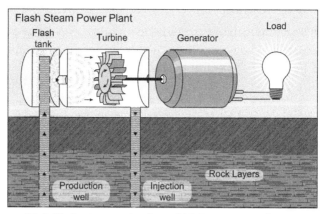

Figure 12.4 *Flash steam technology.* *(U.S. Department of Energy 2012).*

exchanger and used to power a generator. The water, which never comes into direct contact with the working fluid, is then injected back into the ground to be reheated.

The choice of which design to use is determined by the resource. If the water comes out of the well as steam, it can be used directly, as in the first design. If it is hot water of a high enough temperature, a flash system can be used; otherwise it must go through a heat exchanger. Since there are more hot water resources than pure steam or high-temperature water sources, there is more growth potential in the heat exchanger design.

Geothermal reservoirs of low-to moderate-temperature water—68°F to 302°F (20°C to 150°C)—provide direct heat for residential, industrial, and commercial uses. This resource is widespread in the United States, and is used to heat homes and offices, commercial greenhouses, fish farms, food processing facilities, gold mining operations, and a variety of other applications. In addition, spent fluids from geothermal electric plants can be subsequently used for direct use applications in so-called "cascaded"operation.

Direct use of geothermal energy in homes and commercial operations is much less expensive than using traditional fuels. Savings can be as much as 80% over fossil fuels. Direct use is also very clean, producing only a small percentage (and in many cases none) of the air pollutants emitted by burning fossil fuels.

In the geothermal industry, low temperature means temperatures of 300 °F (149 °C) or less. Low-temperature geothermal resources are typically used in direct-use applications, such as district heating, greenhouses, fisheries, mineral recovery, and industrial process heating. However, some low-temperature resources can generate electricity using binary cycle electricity generating technology.

Approximately 70 countries made direct use of 270 petajoules (PJ) of geothermal heating in 2004. More than half went for space heating, and another third for heated pools. The remainder supported industrial and agricultural applications. Global installed capacity was 28 GW, but capacity factors tend to be low (30% on average) since heat is mostly needed in winter. The above figures are dominated by 88 PJ of space heating extracted by an estimated 1.3 million geothermal heat pumps with a total capacity of 15 GW. Heat pumps for home heating are the fastest-growing means of exploiting geothermal energy, with a global annual growth rate of 30% in energy production.

Direct heating is far more efficient than electricity generation and places less demanding temperature requirements on the heat resource. Heat may come from co-generation via a geothermal electrical plant or from smaller wells or heat exchangers buried in shallow ground. As a result, geothermal heating is economic at many more sites than geothermal electricity generation. Where natural hot springs are available, the heated water can be piped directly into radiators. If the ground is hot but dry, earth tubes or downhole heat exchangers can collect the heat. But even in areas where the ground is colder than room temperature, heat can still be extracted with a geothermal heat pump more cost-effectively and cleanly than by conventional furnaces. These devices draw on much shallower and colder resources than traditional geothermal techniques, and they frequently combine a variety of functions, including air conditioning, seasonal energy storage, solar energy collection, and electric heating. Geothermal heat pumps can be used for space heating essentially anywhere.

Geothermal heat supports many applications. District heating applications use networks of piped hot water to heat many buildings across entire communities. In Reykjavík, Iceland, spent water from the district heating system is piped below pavement and sidewalks to melt snow. Geothermal desalination has been demonstrated.

The largest geothermal system now in operation is a steam-driven plant in an area called the Geysers, north of San Francisco, California (Figure 12.5). Despite the name, there are actually no geysers there, and the heat that is used for energy is all steam, not hot water. Although the area was known for its hot springs as far back as the mid-1800s, the first well for power production was drilled in 1924. Deeper wells were drilled in the 1950s, but real development didn't occur until the 1970s and 1980s. By 1990, 26 power

Figure 12.5 *A geothermal power plant at The Geysers near Santa Rosa, California. (Photo: Julie Donnelly-Nolan, U.S. Geological Survey 2009).* (See color plate 36.)

plants had been built, for a capacity of more than 2,000 MW (Union of Concerned Scientists 2009).

REFERENCES

Bertani, Ruggero, Thain, Ian, July 2002. Geothermal Power Generating Plant CO2 Emission Survey, IGA News (International Geothermal Association) (49): 1–3, retrieved 2009-05-13.

Cothran, Helen, March 1, 2002. Energy Alternatives. Greenhaven Press. 220 pages.

Fridleifsson, Ingvar B., Bertani, Ruggero, Huenges, Ernst, Lund, John W., Ragnarsson, Arni, Rybach, Ladislaus, 2008-02-11. In: Hohmeyer, O., Trittin, T. (Eds.), The possible role and contribution of geothermal energy to the mitigation of climate change. Luebeck, Germany, pp. 59–80, retrieved 2009-04-06.

Glassley, William E, 2010. Geothermal Energy: Renewable Energy and the Environment, CRC Press.

Geothermal Energy Association. Geothermal Energy: International Market Update May 2010, p. 4–6,7.

Khan, M. Ali, 2007. (pdf), The Geysers Geothermal Field, an Injection Success Story. Annual Forum of the Groundwater Protection Council, retrieved 2010-01-25.

Nemzer, J., 2012. Geothermal heating and cooling. Cambridge University Press, Cambridge, England, UK, pp. 136, 137, ISBN 978-0-521-66624-4.

Our Energy, 2012. Geothermal energy facts [WWW] Our Energy. Available from: http://www.our-energy.com/energy_facts/geothermal_energy_facts.html.

Rybach, Ladislaus, September 2007. Geothermal Sustainability. In: Geo-Heat Centre Quarterly Bulletin. Oregon Institute of Technology, Klamath Falls, Oregon 28(3): 2–7, ISSN 0276-1084, retrieved 2009-05-09.

Turcotte, D.L., Schubert, G., 2002. "4", Geodynamics, second ed. Cambridge University Press, Cambridge, England, UK, pp. 136–137, ISBN 978-0-521-66624-4.

Union of Concerned Scientists, 2009. How geothermal energy works [WWW] Clean Energy. Available from: http://www.ucsusa.org/clean_energy/technology_and_impacts/energy_technologies/how-geothermal-energy-works.html.

U.S. Department of Energy (DOE), 2012. Geothermal technologies program [WWW] Energy Efficiency & Renewable Energy. Available from: http://www1.eere.energy.gov/geothermal/powerplants.html.

CHAPTER THIRTEEN

Energy Storage, Smart Grids and Electric Vehicles

13.1. ENERGY STORAGE

Energy storage is accomplished by devices or physical media that store some form of energy to perform some useful operation at a later time. A device that stores energy is sometimes called an *accumulator* (Figure 13.1).

All forms of energy are either potential energy (e.g., chemical, gravitational, electrical energy, etc.) or kinetic energy (e.g., thermal energy) (Wagner 2007).

The general method and specific techniques for storing energy are derived from some primary source in a form convenient for use at a later time when a specific energy demand is to be met, often in a different location. In the past, energy storage on a large scale had been limited to storage of fuels. Now, applications like hydroelectric dams store energy in

Figure 13.1 *District heating accumulation tower from Theiss near Krems an der Donau in Lower Austria with a thermal capacity of 2 GWh. (Ulrichulrich 2010).*

a reservoir (gravitational energy), or ice storage tanks store ice (thermal energy) at night to meet peak demand for cooling. On a smaller scale, electric energy is stored in batteries (chemical energy) that power automobile starters and a great variety of portable appliances. In the future, energy storage in many forms is expected to play an increasingly important role in shifting patterns of energy consumption away from scarce to more abundant and renewable primary resources.

An example of growing importance is the storage of electric energy generated during the day by solar, wind energy or other renewable power plants to meet peak electric loads during daytime periods. This is achieved by pumped hydroelectric storage, that is, pumping water from a lower to a higher reservoir and reversing this process at night, with the pump then being used as a turbine and the motor as a generator.

Off-peak electric energy can also be converted into mechanical energy by pumping air into a suitable cavern where it is stored at pressures up to 80 atm (8 megapascals). Turbines and generators can then be driven by the air when it is heated and expanded (The Free Dictionary 2002).

13.1.1. Storage Methods

Energy storage methods can be generally categorized as follows:
- Chemical
  - Hydrogen
  - Biofuels
  - Liquid nitrogen
  - Oxyhydrogen
  - Hydrogen peroxide
- Biological
  - Starch
  - Glycogen
- Electrochemical
  - Batteries
  - Flow batteries
  - Fuel cells
- Electrical
  - Capacitor
  - Supercapacitor
  - Superconducting magnetic energy storage (SMES)
- Mechanical
  - Compressed air energy storage (CAES)

- • Flywheel energy storage
- • Hydraulic accumulator
- • Hydroelectric energy storage
- • Spring
- • Gravitational potential energy (device)
- Thermal
  - • Ice storage
  - • Molten salt
  - • Cryogenic liquid air or nitrogen
  - • Seasonal thermal store
  - • Solar pond
  - • Hot bricks
  - • Graphite accumulator very high temperature
  - • Steam accumulator
  - • Fireless locomotive
  - • Eutectic system
- Fuel Conservation storage (Wagner 2007)

Electrolysis has been around for many decades and is widely used for the production of oxygen and hydrogen in the chemical industry, paper industry, hospitals and for welding. For energy storage, hydrogen is still in the early stage of development. Initial costs are high due to high pressure and diffusion of hydrogen, and conventional gas storage equipment is not suitable. Losses of conversion in the process from electricity back to electricity may be 65–80% accumulated by losses in rectifier, electrolyzer, compression, transmission and the fuel cell (QuantumSphere Inc. 2006).

There are several commercially viable energy storage systems that are being developed for hybrid electric vehicles (HEVs) on the market today. The types of devices that hold the most promise to solve the energy storage problems are batteries, flywheels, and ultracapacitors. As shown in Figure 13.2, both gasoline and hydrogen have a higher specific energy than the rest of these electrical storage devices (Fuel Cells 2000, 2008).

An advantage of HEVs is that they can use the high specific energy of liquid or gaseous fuels to provide the vehicle with long-range capabilities. Conversely, the HEV can use the high specific power of electrical energy storage to provide the peak power requirements.

Batteries for storage of electricity are widely used in many applications. For electric cars a new generation of lithium batteries are being developed in many industrialized countries; they are expected gradually to be available for large-scale storage as well. For the hundreds of thousands of stand-alone

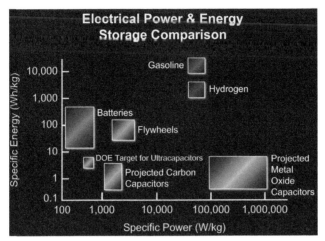

Figure 13.2 *Plot of energy versus power for various energy storage devices. (Diagram Provided Courtesy of National Renewable Energy Laboratory).* (See color plate 37.)

PV installations in the developing countries, conventional car and truck batteries are used as well as batteries designed for frequent deep-cycling. For regional and national supply structures such batteries will have a limited application. Research is being conducted on harnessing the quantum effects of nanoscale capacitors to create digital quantum batteries. Although this technology is still in the experimental stage, it theoretically has the potential to provide dramatic increases in energy storage capacity (Doughty, et al. 2010). A flywheel (Figure 13.3) energy storage system draws electrical energy from a primary source, and stores it in a high-density rotating flywheel. It is effectively a kinetic battery, spinning at very high speeds (>20,000 rpm) to store energy that is instantly available when needed. Upon power loss, the motor driving the flywheel acts as a generator, supplying power to the customer load (Electrical Storage 2012).

Another possible technology is ultra-capacitors. These devices work by accumulating and separating unlike charges. The positive abilities lie in the fact that they have no moving parts, as well as the fact that the number of times they can be cycled through their charge–discharge cycle is very high. The energy density of supercapacitors is 100 times higher than in normal capacitors and the power density is 10 times higher than in normal batteries allowing their use in portable electronics, electric vehicles and for storing energy generated from renewable sources such as wind and solar power (Wagner 2008).

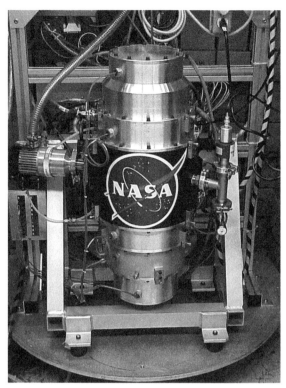

Figure 13.3 *G2 Flywheel Module, NASA. (Image: NASA Aerospace Flywheel Technology Program).*

Electrochemical devices called fuel cells were invented about the same time as the battery in the 19th century. However, for many reasons, fuel cells were not well-developed until the advent of manned spaceflight (such as the Gemini program in the U.S.) when lightweight, non-thermal (and therefore efficient) sources of electricity were required in spacecraft. Fuel cell development has increased in recent years due to an attempt to increase conversion efficiency of chemical energy stored in hydrocarbon or hydrogen fuels into electricity (Wagner 2007).

Several other technologies have also been investigated: compressed air storage that can be pumped into underground caverns and abandoned mines (Wild 2010), and a method used at the Solar Project and the Solar Tres Power Tower which uses molten salt to store solar power and then dispatch that power as needed. The system pumps molten salt through a tower heated by the sun's rays. Insulated containers store the hot salt solution and, when needed, water is then used to create steam that is fed to turbines to generate electricity It can be used alone or combined with wind energy in utility size installations of 50 MW or bigger as it is being demonstrated in southern Spain and in the USA. With operational temperatures up to 400°C the storage medium can produce steam for conventional steam turbines combined with production of electricity. The process heat can be distributed by a district heating network for heating and for cooling by absorption chillers (NREL 2011).

Compressed Air Energy Storage (CAES) is a way to store energy generated at one time for use at another time and has been in operational use for several years in the USA and Germany. Off-peak (low-cost) electrical power compresses air into an underground air-storage "vessel" (Figure 13.4) and later the air feeds a gas-fired turbine generator complex to generate electricity during on-peak (high-price) times (Wild 2010).

The excess fluctuating electricity is used for the compression of atmospheric air into deep underground caverns of a type similar to natural gas storage. At the time of consumption the process is reversed and the air drives a conventional type turbine that, instead of natural gas or steam, uses compressed air which is connected to a generator. During compression heat is produced while the reversal process occurs with decompression and the air expands so that the system can deliver cooled air. The electric efficiency is around 50%; the overall efficiency can be improved if the heating and

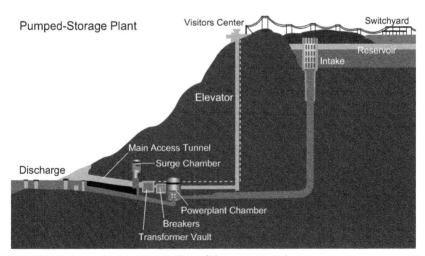

Figure 13.4 *Conceptual representation of the compressed-air energy storage concept.* *Tennessee Valley Authority (TVA) 2004. http://www.tva.gov/power/pumpstorart.htm* (See color plate 38.)

cooling potential is utilized. A similar concept uses wind-powered air compressors (Pockley 2008).

Water pumped storages are installed in many countries to compensate for fluctuations in the demand for power (Figure 13.5). Pump storages have dual purposes. A pumped-storage plant is designed with two reservoirs—upper and lower. Like every other hydroelectric plant, a pumped-storage plant generates electricity by allowing water to fall through a turbine generator. But unlike conventional hydroelectric plants, once the pumped-storage plant generates electricity, it can then pump that water from its lower reservoir back to the upper reservoir. This is done during the off-peak hours, using electricity from another source to run the plant's pumps, in effect, "storing" that off-peak electricity (Duke Energy 2012). Their general application is limited by the topography; in Europe most potential sites for pump storage have already been developed.

Other storage solutions can be mentioned as well. *Molten salt* is used for concentrated solar power storage. It can be used alone or combined with wind energy in utility size installations of 50 MW or bigger, as it is being

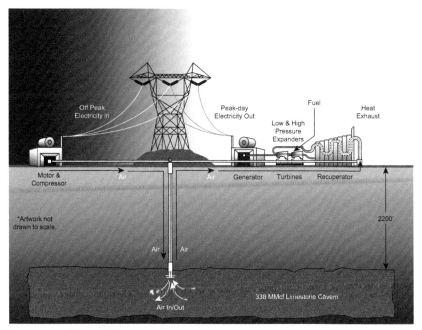

Figure 13.5 *Pumped Storage diagram at TVA's Raccoon Mountain, United States (Sandia National Laboratories) http://www.sandia.gov/media/NewsRel/NR2001/images/jpg/minebw.jpg.*

demonstrated in southern Spain and in the USA. With operational temperatures up to 400°C the storage medium can produce steam for conventional steam turbines combined with production of electricity. The process heat can be distributed by a district heating network for heating and for cooling by absorption chillers (Mancini 2006).

In Denmark several hundred hot water storage tanks are installed at local CHP; sizes range from 10 m^3 up to 30.000 m^3. The dimensioning criteria is often to cover the CHP station's need for supply to the district heating network during the power low peak period during the weekends.

Energy storage has critical roles to play in securing our energy future including:

- serving as an "electricity reserve" much like the national Petroleum Reserve
- stabilizing electricity markets
- stabilizing the transmission and distribution grid
- enabling more efficient use of existing generation assets
- making renewable energy economically viable (Maegaard 2011)

13.1.2. Technologies for Up- and Down-Regulation

Many renewable energy sources (most notably solar and wind) produce intermittent power. Wherever intermittent power sources reach high levels of grid penetration, energy storage becomes one option to provide reliable energy supplies. Other options include recourse to peaking power plants, methane storage (excess renewable electricity to hydrogen via electrolysis, combine with CO2 [low to neutral CO2 system] to produce methane [synthetic natural gas sabatier process] with stockage in the natural gas network) and smart grids with advanced energy demand management. The latter involves bringing "prices to devices"—i.e., making electrical equipment and appliances able to adjust their operation to seek the lowest spot price of electricity. On a grid with a high penetration of renewables, low spot prices would correspond to times of high availability of wind and/or sunshine.

For economic evaluation of large-scale applications, like pumped hydro storage and compressed air, the potential benefits that can be accounted for are avoidance of wind curtailment, grid congestion avoidance, price arbitrage and carbon-free energy delivery.

Solar and wind alone cannot ensure a continuous supply of electricity. In a future supply scenario dominated by fluctuating energy forms based in full on renewables, three typical situations may occur to meet the actual demand for electricity:

a. Solar and wind produce too *much* power; down-regulation is required and some of the excess power can be converted, stored or exported

b. Solar and wind produce too *little* power; up-regulation is required and the stored energy will be used and power imported

c. Solar and wind produce *no* power; storageable supply and import is required and will cover the total demand

Wind power and PV power are cornerstones in future renewable integrated energy supply structures. They are, however, fluctuating, which causes a need for adaptation by consumers, and back-up from other supply solutions or storage.

A number of storage solutions are available. They are of very different character regarding technology, medium and cost (Figure 13.6). Flexibility and response time are important requirements that the various types of energy storage will meet differently, to match satisfactorily with the integrated supply of power, heat and cooling.

A storage solution that converts the residual power from wind and solar may be chemical, gravity, heat, compression, etc., and may be suitable for either conversion back to electricity or heating/cooling. The need for storage may be for seconds, minutes, hours or a few days. Seasonal storage is well developed for hot water, but is rare for big-scale electricity storage (Maegaard 2011).

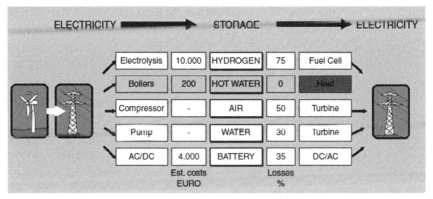

Figure 13.6 *Storage of electricity.* *(Maegaard, Power 2004).*

13.2. SMART GRIDS

13.2.1. Definition and Importance

A *smart grid* is a digitally enabled electrical grid that gathers, distributes, and acts on information about the behavior of all participants (suppliers and consumers) in order to improve the efficiency, importance, reliability, economics, and sustainability of electricity services (U.S. DOE 2012). The term *smart grid* is an umbrella term that covers modernization of both the transmission and distribution grids. The modernization is directed at a disparate set of goals including facilitating greater competition between providers, enabling greater use of variable energy sources, establishing the automation and monitoring capabilities needed for bulk transmission at cross continent distances, and enabling the use of market forces to drive energy conservation.

Another element of fault tolerance of traditional and smart grids is decentralized power generation. Distributed generation allows individual consumers to generate power onsite, using whatever generation method they find appropriate. This allows individual loads to tailor their generation directly to their load, making them independent from grid power failures. Classic grids were designed for one-way flow of electricity, but if a local sub-network generates more power than it is consuming, the reverse flow can raise safety and reliability issues. A smart grid can manage these situations, but utilities routinely manage this type of situation in the existing grid.

An electrical grid is not a single entity but an aggregate of multiple networks and multiple power generation companies with multiple operators employing varying levels of communication and coordination, most of which is manually controlled. Smart grids increase the connectivity, automation and coordination between these suppliers, consumers and networks that perform either long-distance transmission or local distribution tasks.

- Transmission networks move electricity in bulk over medium to long distances, are actively managed, and generally operate from 345kV to 800kV over AC and DC lines.
- Local networks traditionally moved power in one direction, "distributing" the bulk power to consumers and businesses via lines operating at 132kV and lower.

This paradigm is changing as businesses and homes begin generating more wind and solar electricity, enabling them to sell surplus energy back to their utilities. Modernization is necessary for energy consumption efficiency and real-time management of power flows and to provide the bidirectional metering needed to compensate local producers of power. Although transmission networks are already controlled in real time, many in the US and European countries are antiquated by world standards, and unable to handle modern challenges such as those posed by the intermittent nature of alternative electricity generation, or continental scale bulk energy transmission (Kaplan 2009).

13.2.2. U.S. Strategy

Support for smart grids became federal policy in the USA with passage of the Energy Independence and Security Act of 2007. The law, Title XIII, sets out $100 million in funding per fiscal year from 2008–2012, establishes a matching program to states, utilities and consumers to build smart grid capabilities, and creates a Grid Modernization Commission to assess the benefits of demand response and to recommend needed protocol standards. The Energy Independence and Security Act of 2007 directs the National Institute of Standards and Technology to coordinate the development of smart grid standards, which FERC would then promulgate through official rulemakings.

Smart grids received further support with the passage of the American Recovery and Reinvestment Act of 2009, which set aside $11 billion for the creation of a smart grid (Wikipedia 2011). Figure 13.7 shows a smart grid definition from the deployment plan of Pacific Gas & Electric (PG&E 2011), one of the largest combined natural gas and electric utilities in the United States.

Section 1301 of Title XIII (2007) reads:

It is the policy of the United States to support the modernization of the Nation's electricity transmission and distribution system to maintain a reliable and secure electricity infrastructure that can meet future demand growth and to achieve each of the following, which together characterize a Smart Grid.

(1) Increased use of digital information and controls technology to improve reliability, security, and efficiency of the electric grid.

(2) Dynamic optimization of grid operations and resources, with full cyber-security.

(3) Deployment and integration of distributed resources and generation, including renewable resources.

(4) Development and incorporation of demand response, demand-side resources, and energy-efficiency resources.

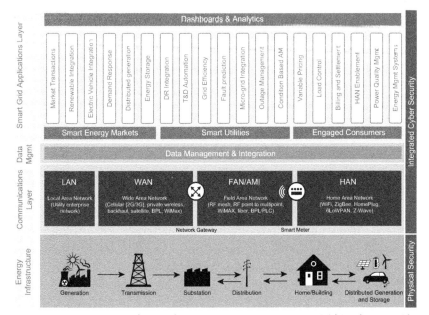

Figure 13.7 *Smart Grid Definition from PG&E's June 2011 Smart Grid Deployment Plan.*
(PG&E 2011). (See color plate 39.)

(5) Deployment of "smart" technologies (real-time, automated, interactive technologies that optimize the physical operation of appliances and consumer devices) for metering, communications concerning grid operations and status, and distribution automation.

(6) Integration of "smart" appliances and consumer devices.

(7) Deployment and integration of advanced electricity storage and peak-shaving technologies, including plug-in electric and hybrid electric vehicles, and thermal-storage air conditioning.

(8) Provision to consumers of timely information and control options.

(9) Development of standards for communication and interoperability of appliances and equipment connected to the electric grid, including the infrastructure serving the grid.

(10) Identification and lowering of unreasonable or unnecessary barriers to adoption of smart grid technologies, practices, and services.

Section 1304 of Title XIII reads in part:

In consultation with the Federal Energy Regulatory Commission and other appropriate agencies, electric utilities, the States, and other stakeholders, shall carry out a program:

(1) to develop advanced techniques for measuring peak load reductions and energy-efficiency savings from smart metering, demand response, distributed generation, and electricity storage systems;

(2) to investigate means for demand response, distributed generation, and storage to provide ancillary services;

(3) to conduct research to advance the use of wide-area measurement and control networks, including data mining, visualization, advanced computing, and secure and dependable communications in a highly distributed environment;

(4) to test new reliability technologies, including those concerning communications network capabilities, in a grid control room environment against a representative set of local outage and wide area blackout scenarios;

(5) to identify communications network capacity needed to implement advanced technologies;

(6) to investigate the feasibility of a transition to time-of-use and real-time electricity pricing;

(7) to develop algorithms for use in electric transmission system software applications;

(8) to promote the use of underutilized electricity generation capacity in any substitution of electricity for liquid fuels in the transportation system of the United States; and

(9) in consultation with the Federal Energy Regulatory Commission, to propose interconnection protocols to enable electric utilities to access electricity stored in vehicles to help meet peak demand loads.

13.2.3. European Strategy

The Smart Grids Platform was started by the European Commission Directorate General for Research of the European Commission in 2005. The **European Technology Platform for Electricity Networks of the Future**, also called Smart Grids ETP, is the key European forum for the crystallization of policy and technology research and development pathways for the Smart Grids sector, as well as the linking glue between EU-level related initiatives.

Development of smart grid technologies is part of the European Technology Platform (ETP) initiative and is called the Smart Grids Technology platform. Its aim is to formulate and promote a vision for the development of European electricity networks looking toward 2020 and beyond. The

European Technology Platform (ETP) for the Electricity Networks of the Future (Smart Grids) is a European Commission initiative that aims at boosting the competitive situation of the European Union in the field of electricity networks, especially smart power grids. The ETP represents all European stakeholders. The establishment of an ETP in this field was for the first time suggested by the industrial stakeholders and the research community at the first International Conference on the Integration of Renewable Energy Sources and Distributed Energy Resources, which was held in December 2004 (European Commission 2006).

In the so-called E-Energy projects several German utilities are creating the first nucleolus in six independent model regions. A technology competition identified these model regions to carry out research and development activities with the main objective of creating an "Internet of Energy" (Figure 13.8) (Federation of German Industries 2010).

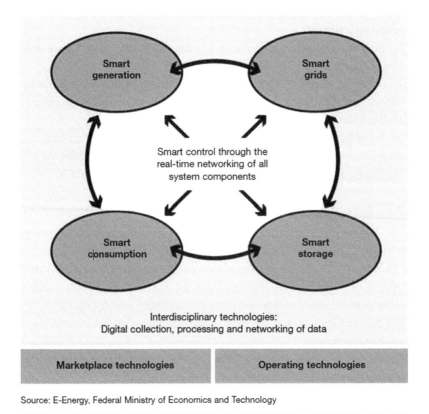

Source: E-Energy, Federal Ministry of Economics and Technology

Figure 13.8 *The Internet of Energy Integrates all the elements in the energy supply chain to create an interactive system. (Federation of German Industries 2010).*

13.2.4. Korean Version

The Korean government has launched a $65 million pilot program on Jeju Island with major players in the industry. The program consists of a fully integrated Smart Grid System for 6000 households; wind farms and four distribution lines are included in the pilot program. This demonstrates the extent of Korea's commitment toward an environmentally viable future.

Korea plans to slash overall energy consumption by 3% and cut down total electric energy consumption by 10% before 2030 (Figure 13.9). The government also plans to reduce greenhouse gas emissions by 41 million tons by this time. The government has announced that it will undertake a nation-wide smart grid implementation by 2030 (Korea Smart Grid Institute [KSGI] 2010).

13.3. ELECTRIC VEHICLES

13.3.1. Current Developments

During the last few decades, environmental impact of the petroleum-based transportation infrastructure, along with peak oil, has led to renewed interest

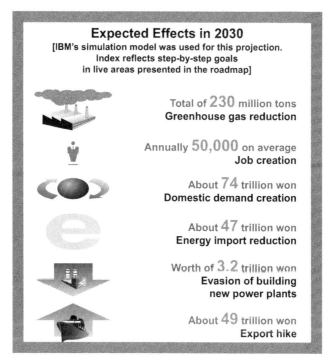

Expected Effects in 2030
[IBM's simulation model was used for this projection. Index reflects step-by-step goals in live areas presented in the roadmap]

Total of 230 million tons
Greenhouse gas reduction

Annually 50,000 on average
Job creation

About 74 trillion won
Domestic demand creation

About 47 trillion won
Energy import reduction

Worth of 3.2 trillion won
Evasion of building new power plants

About 49 trillion won
Export hike

Figure 13.9 *Korea's Smart Grid expected result in 2030. Korea Smart Grid Institute [KSGI] 2010.*

in an electric transportation infrastructure (Eberle 2010). Electric vehicles differ from fossil fuel-powered vehicles in that the electricity they consume is generated from a wide range of renewable sources. However it is generated, this energy is then transmitted to the vehicle through use of overhead lines, wireless energy transfer such as inductive charging, or a direct connection through an electrical cable. The electricity may then be stored on board the vehicle using a battery, flywheel, or supercapacitors. Vehicles making use of engines working on the principle of combustion can usually only derive their energy from a single or a few sources, usually non-renewable fossil fuels. A key advantage of electric or hybrid electric vehicles (Figure 13.10) is regenerative braking and suspension; their ability to recover energy normally lost during braking as electricity to be restored to the on-board battery (Levant Power Corp 2011).

Since electric vehicles can be plugged into the electric grid when not in use, there is a potential for battery-powered vehicles to even out the demand for electricity by feeding electricity *into* the grid from their batteries during peak use periods (such as mid-afternoon air conditioning use) while doing most of their charging at night, when there is unused generating capacity (Pacific Gas and Electric Company 2007). This Vehicle to Grid (V2G) connection has the potential to reduce the need for new power plants.

Furthermore, our current electricity infrastructure may need to cope with increasing shares of variable-output power sources such as windmills and PV solar panels. This variability could be addressed by adjusting the speed at which EV batteries are charged, or possibly even discharged.

Figure 13.10 *Plug-in hybrid car.*

Some concepts see battery exchanges and battery charging stations, much like gas/petrol stations today. Clearly these will require enormous storage and charging potentials, which could be manipulated to vary the rate of charging, and to output power during shortage periods, much as diesel generators are used for short periods to stabilize some national grids (Andrews 2006).

With electric cars gaining in popularity, AEP Ohio and Walmart premiered the region's first free, public electric-vehicle charging station at the Walmart Supercenter/Sam's Club at 3900 Morse Road (Figure 13.11). The Blink® charging station was developed by San Francisco-based ECOtality, Inc., a provider of clean electric transportation and storage technologies. The charging station features two Blink Pedestal units that allow two drivers to charge their electric vehicles simultaneously.

According to AEP, by 2015, there could be a million PEVs on the road, which is why AEP has launched the demonstration project to determine the

Figure 13.11 *Ohio's first electric car charging station.* *(New Albany Innovation Exchange 2011).*

best way to integrate plug-in electric vehicles into its electricity distribution system. To analyze the new demand, up to ten people will drive a PEV as part of their daily routine to assess the amount of power needed for basic travel, effects of PEV charging during peak electricity periods and the quality of charging technology.

The charging station is part of AEP Ohio's gridSMART® initiative to promote energy efficiency and Walmart's corporate "Sustainability 360" effort. The gridSMART® initiative is designed to provide customers greater energy control and improve electric distribution service and performance through rebates and cost incentives, plus tips, tools and technologies (New Albany Innovation Exchange 2011).

13.3.2. Future Developments

The ability to power wireless sensor nodes from harvested energy sources allows embedded designers to offer systems with significantly reduced cost of ownership for the end-user as well as benefits to the environment. The cost of replacing batteries housed in out-of-the-way sensor node locations can be

Figure 13.12 *Fast charging station for future electric vehicles.* *(Milano Medien GmbH / Siemens AG 2009).*

quite significant. These wireless sensor nodes, for example, can be embedded in structures, such as buildings or bridges, or even buried underground. The three key enabling technologies needed to create self-sustaining wireless sensor nodes are readily available today: cost-effective energy harvester devices, small and efficient energy storage devices and single-chip ultra-low-power wireless MCUs. Wireless sensor nodes powered by harvested energy sources will soon become commercially viable and commonplace technologies used in our homes, offices, factories and infrastructure (Silicon Laboratories 2011).

The high cost of electric cars is one of the most common reasons consumers cite for choosing not to own one. For those people, a pair of green tech startups plan to implement a program that would enable EV owners to offset some of the costs by selling electric storage services to help stabilize the electricity grid (Nguyen 2011).

Figures 13.12, 13.13, and 13.14 show plans for some future technologies related to electric vehicles.

Figure 13.13 *Fast charging station for future electric vehicles embedded in public places such as shopping malls. (Milano Medien GmbH / Siemens AG 2009).* (See color plate 40.)

Figure 13.14 *Autonomous solar power/electricity/combined heat and power at microgrid of a family house including electric car charging facilities.* *(Milano Medien GmbH / Siemens AG 2009).* (See color plate 41.)

REFERENCES

Andrews, David, 2006. Senior Technical Consultant, Biwater Energy. January 2006. A talk originally given by as the Energy Manager at Wessex Water at an Open University Conference on Intermittency 24th National Grid's use of Emergency. Diesel Standby Generator's in dealing with grid intermittency and variability. Potential Contribution in assisting renewables.

Doughty, Daniel, Butler, H., Paul, C., Akhil, Abbas A., Clark, Nancy, H., Boyes, John D., 2010. Batteries for Large-Scale Stationary Electrical Energy Storage. The Electrochemical Society Interface. http://www.electrochem.org/dl/interface/fal/fal10/fal10_p049-053.pdf.

Duke Energy, 2012. Generating Electricity with Pumped-Storage Hydro. http://www.duke-energy.com/about-energy/generating-electricity/pumped-storage-faq.asp.

Eberle, Ulrich; von Helmolt, Rittmar (2010-05-14). "Sustainable transportation based on electric vehicle concepts: a brief overview". Royal Society of Chemistry, http://pubs.rsc.org/en/Content/ArticleLanding/2010/EE/c001674h.

Electrical Storage 2012. http://www.microchap.info/electrical_storage.htm.

Energy Independence and Security Act, U.S.C. Title XIII Sec. 1301 (2007). Retrieved from http://energy.gov/sites/prod/files/oeprod/DocumentsandMedia/EISA_Title_XIII_Smart_Grid.pdf.

European Commission, 2006. European SmartGrids Technology Platform: Vision and Strategy for Europe's Electricity Networks of the Future Luxembourg, EUR 22040.

Federation of German Industries/BDI Initiative Internet of Energy, 2010. Internet of Energy ICT for Energy Markets of the Future: The Energy Industry on the Way to the Internet

Age (translation of the brochure "Internet der Energie – IKT für Energiemärkte der Zukunft" published in Germany, December 2008, to which information about the German government's E-Energy model projects has been added). Available from http://www.bdi. eu/bdi_english/download_content/Marketing/Brochure_Internet_of_Energy.pdf.

Fuel Cells, 2008. Fuel Cell Basics, 2000. http://www.fuelcells.org/basics/types.html.

Kaplan, S.M., 2009. Smart Grid. Electrical Power Transmission: Background and Policy Issues. The Government Series, 1–42. Capital.Net.

Korea Smart Grid Institute (KSGI), 2010. Korea's Jeju Smart Grid [WWW]. Available from http://www.smartgrid.or.kr/10eng3-3.php.

Levant Power Corp, 2011. Revolutionary GenShock Technology. http://www.levantpower.com/technology.html.

Maegaard, Preben, June 2011. Integrated Systems to Reduce Global Warming 2011, Springer Science+Business Media. LLC 22 PDF.

Maegaard, P., 2004. Wind energy development and application prospects of non-grid-connected wind power 2004, World Wind Energy Institute. World Renewable Energy Committee, Nordic Folkecenter for Renewable Energy, Denmark.

Mancini, Tom, 2006. Advantages of Using Molten Salt. Sandia National Laboratories. Archived from the original on 2011-07-14,. http://www.webcitation.org/60AE7heEZ.

New Albany Innovation Exchange. Ohio's First Electric Car Charging Station, posted on September 15, 2011. http://www.innovatenewalbany.org/business/ohio%E2%80%99s-first-electric-car-charging-station/.

Nguyen, T.C., 2011. Electric cars can now earn money for owners. September 29, 2011. Smart planet. Available from http://www.smartplanet.com/blog/thinking-tech/electric-cars-can-now-earn-money-for-owners/8756.

Pacific Gas and Electric Company, 2007. First vehicle-to-grid demonstration. http://seekingalpha.com/article/31992-pacific-gas-and-electric-demonstrates-vehicle-to-grid-technology.

Pacific Gas & Electric (PG&E), 2011. Smart Grid Definition. Smart grid deployment plan. Available from http://www.neuralenergy.info/2009/09/pg-e.html.

Pockley, Simon, 19/05/2008. Compressed Air Energy Storage (CAES), Prepared for Intro. to Renewable Energy (PDF).

QuantumSphere Inc, 2006. Highly Efficient Hydrogen Generation via Water Electrolysis Using Nanometal Electrodes. http://www.qsinano.com/white_papers/2006_09_15.pdf.

U.S. Dept. of Energy (DOE), October 2011. "Gemasolar Thermosolar Plant". Concentrating Solar Power Projects. National Renewable Energy Laboratory (NREL) 24.

U.S. Department of Energy. "Smart Grid / Department of Energy". Retrieved 2012-06-18.

Silicon Laboratories, 2011. The energy harvesting tipping point for wireless sensor applications [WWW]. White paper available from http://www.eetimes.com/electrical-engineers/education-training/tech-papers/4217176/The-Energy-Harvesting-Tipping-Point-for-Wireless-Sensor-Applications.

Tennessee Valley Authority (TVA), 2004. Energy [WWW]. Available from http://www.tva.gov/power/pumpstorart.htm.

The Free Dictionary, 2002. McGraw-Hill Concise Encyclopedia of Engineering. http://encyclopedia2.thefreedictionary.com/Energy+storage Energy storage.

Wagner, Leonard, January 2008. Nanotechnology in the clean tech sector. Research report. http://www.moraassociates.com/publications/0801%20Nanotechnology.pdf.

Wagner, Leonard, December 2007. Overview of energy storage methods. Research report. http://www.moraassociates.com/publications/0712%20Energy%20storage.pdf.

Wild, Matthew, L., July 28, 2010. Wind Drives Growing Use of Batteries. New York Times, pp. B1.

CHAPTER FOURTEEN

Current Distributed Renewable Energy Rural and Urban Communities

This chapter reproduces some selected implemented and planned renewable energy communities: rural, remote and urban. It should be pointed out that many more communities have been established across the continents.

14.1. RURAL COMMUNITY JÜHNDE

The community project Jühnde (Figure 14.1) was inspired by the IEC concept and other institutions. The village of Jühnde (750 inhabitants) is located in Southern Lower Saxony in the middle of Germany. It was started in 2001 to become a "Bio Energy Village." With one third of funds from the German Ministry BMELV and Lower Saxony, it was possible to invest in such a project. The main emphasis is that the whole village is involved and more than 70 percent of the households are connected to the hot water grid. The aim of the project is to convert biological material into electrical power and heat. A block-type thermal power station (or heat and power generator) run by biogas is now realized. For additional heating during winter a wood hogged heating system is implemented (IZNE, 2007).

14.1.1. The Energy Production Process

Under anaerobic conditions, micro-organisms engage in enzymatic digestion to create biogas. Biogas is obtained during the fermentation process of liquid manure and plant silage in an anaerobic digestion plant. The combustion of biogas in a combined heat and power generator (CHP) generates enough electricity for the entire village.

Biogas also generates heat as a byproduct. This heat is mainly used to heat homes and other living spaces, replacing the conventional fossil fuels, oil and natural gas (Figure 14.2). A smaller portion of the generated heat is required to fuel the digestion process described above. The amount of heat generated cannot cover the high demand during winter months in Germany. During

Figure 14.1 *Bioenergy Village Jühnde, Germany, with the energy generation installations. (Nachwachsende Rohstoffe, El Bassam, 2010).*

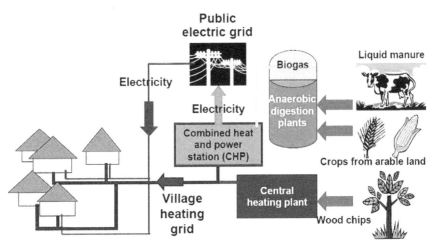

Figure 14.2 *Technical concept of the project. (Ruwisch, V. & Sauer, B. 2007 and IZNE (2005).*

this period, an additional heating plant fueled with wood chips is required. Rarely, on extremely cold days, peak demand necessitates a further boiler fueled by oil or biodegradable diesel.

The distribution of heat energy to the 140 households in Jühnde (750 inhabitants) started in 2005. In 2008 the project produced more than 10,000,000kWh electricity and is saving 3,300 tons of CO_2 annually.

The bioenergy project is contributing toward reduction of CO_2 emissions by 3300 tons/year = 60 percent CO_2 reduction/ capita/year. It has already reached the CO_2-reduction aims of the European Union for 2005.

14.2. WILDPOLDSRIED, THE 100% EMISSIONS FREE TOWN

Germany is paving the way in the field of green energy (Figure 14.3). For the first time ever, renewable energy sources accounted for more than 20 percent of the country's power in the first half of 2011. But there's still a long way to go before Germany is 100% emissions free.

That's where Wildpoldsried comes in. The small town of 2,500 people in southern Germany runs completely on clean energy by using a mix of solar, wind, hydropower, geothermal and biomass (Figure 14.4):

- Five wind turbines total 7500 kW
- Biomass heating plant with 400 KW district heating network
- Four biogas plants with 8 BHWK biogas (cogeneration) total, 1300 kW

Figure 14.3 *Locations of renewable energy communities in Germany. (Source: German American Bioenergy Conference/Syracuse, Eckhard Fangmeier 2009-06-23).*

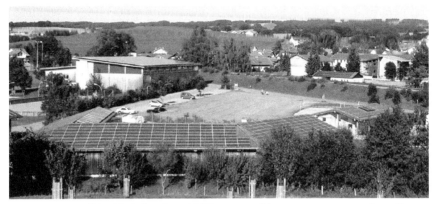

Figure 14.4 *Wildpoldsried, the 100% renewable energy town. (http://www. wildpoldsried.de/).*

- 190 PV systems > 3300 kWp
- 140 Solar thermal systems > 1900 m
- Three hydropower plants sat. 53 kW
- Four geothermal power plants
- Private wooden fireplaces 9000 kW/700 plants

Wildpoldsried generates (17,990 MWh), three times the amount of energy it consumes (5.590 MWh). During winter, the town uses pellets made from renewable wood to heat every house. With 24 million Euro in investments, it is generating 3.6 million Euro in annual revenue.

The town expects to generate in 2012 more than 30,900 MWh—i.e., at least five times the energy needed with an annual revenue of **4,755,932 Euro** (Allen, 2011).

14.3. ROADMAP TO RENEWABLE ENERGY IN REMOTE COMMUNITIES IN AUSTRALIA

In late 2009, the Chief Minister of the Northern Territory (NT) formed the Green Energy Taskforce and charged them with developing a roadmap for the development of the renewable and low emission energy sector and products in the Territory.

In 2009, over 31 million liters of diesel fuel was used to generate electricity for the major remote communities in the Northern Territory, and demand for electricity across these locations was expected to increase by a further 25% over the next three (3) years.

The NT has a strong history of renewable and low emissions energy production in urban and remote areas and the capability to expand the number of renewable and low emission energy systems deployed. Northern Territory organizations with experience and expertise in researching and implementing renewable energy include public sector entities, private sector companies, not-for-profit organizations and academic institutions.

Up to this point, a total of 669 renewable energy generating systems have been funded in the Northern Territory through the Renewable Remote Power Generation Program (RRPGP), an Australian government program that closed in 2009. Many of the systems are small in scale but large-scale renewable energy systems have also been deployed across Yuendumu, Hermannsburg, Lajamanu, Kings Canyon, Bulman and Jilkminggan.

Three new photovoltaic (PV) systems deployed under the RRPGP at Ti Tree, Kalkarindgi and Alpurrurulam in 2011 are more advanced in terms of the diesel-PV hybrid configuration and a high level of PV penetration (i.e., replacement of diesel power with renewable solar power). Lessons can be learned from this project to support future PV deployment and broader expansion of the initiatives proposed.

The renewable energy technologies currently available/deployable in the Northern Territory are PV, wind and solar thermal systems. While PV systems are technically proven and becoming more economic (with falling PV panel prices), the primary issue for solar thermal is that the economic scale is beyond what is being considered for remote communities. The wind resource in the Territory is limited, and although other resources such as geothermal and tidal energy are abundant, the applicable technologies have not yet been commercialized. Bio-fuels have been commercialized to a limited extent.

The Northern Territory Government, through its Climate Change Policy, has committed to replacing diesel as a primary source of power generation to remote towns and communities, developing a green energy industry, reducing greenhouse gas contributions and assist Power and Water Corporation (PWC) to meet its Renewable Energy Target (RET) obligations through local sources of renewable energy.

To identify how this can occur, the Green Energy Taskforce was established by the Chief Minister to provide expert advice (including their Roadmap Report) on strategies, incentives, and pathways to develop renewable and low emission energy and products in the Northern Territory.

The first major task identified in the Terms of Reference for the Taskforce was to develop a proposal for substituting a large component of diesel generation with low emissions and renewable energy in remote communities by 2020 (Figure 14.5).

The Roadmap is, however, broader than just electricity generation, as it also identifies opportunities for skills transfer, training and management in communities, Indigenous economic development, "Closing the Gap of Indigenous Disadvantage" and remote service delivery reform under the "Working Futures" policy, through the development of a low emissions and renewable energy industry in the Northern Territory.

The first component is to roll-out 10MW of PV to 46 remote communities. This first step involved the integration of solar PV capacity into 46 diesel power stations, up to a level that produced significant savings in diesel but did not require expensive modifications to generators or installation of energy storage equipment or infrastructure (e.g., batteries, fly wheels, etc.). These communities do not have, nor are there immediate plans to install, significant renewable energy systems.

A roll-out of 10MW of PV was proposed to reduce diesel use by 17%, supply approximately 30% of peak demand across the selected communities and also meet 7% of PWC's cumulative RET obligation to 2030.

This level of renewable energy penetration was recommended as it did not require expensive storage or modifications to existing power plants and as such represented the least-cost renewable energy option for significant diesel fuel savings. Variability in renewable energy output at this level was within performance thresholds of the existing diesel generation plants.

(Note: The potential for a complementary energy efficiency program was explored later in 2009 by the Taskforce, to further reduce diesel use and maximize benefits from diesel substitution in light of increasing demand over time.)

It was anticipated that this first component would take 3–4 years to implement, with a total cost of $60 million.

The technical and economic viability of other energy sources including pipeline natural gas, liquefied petroleum gas (LPG), bio-diesel, compressed natural gas, liquefied natural gas and very hi-penetration PV will continue to be monitored by existing internal processes.

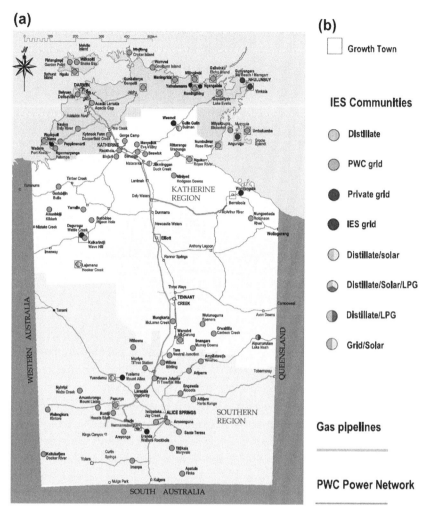

Figure 14.5(a) *Energy sources used by IES PTY LTD to provide essential services to indigenous communities in the NT; (b) Map details.* (*Northern Territory Government Australia: Roadmap to Renewable and Low Emission Energy in Remote Communities 2011*).

However, to be successful in achieving 100% diesel substitution, and to achieve this target both efficiently and effectively, will require the integration of a range of fuel sources and technologies.

The second component is moving toward 100% substitution of diesel generation in NT remote communities. The second component of the Roadmap proposes defining options for a more ambitious target of moving

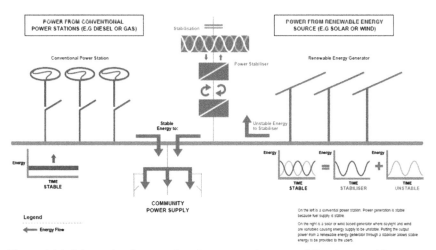

Figure 14.6 *Schematic of conventional power station vs. stabilized renewable energy power station. (Northern Territory Government Australia: Roadmap to Renewable and Low Emission Energy in Remote Communities 2011).*

toward 100% substitution of diesel generation with renewable energy and low emissions fuel in remote communities and will be pursued concurrently with the first component (roll-out of 10MW of renewable energy in 46 remote communities).

Maximization of the effectiveness and potential of renewable energy in off-grid power stations will involve optimization to suit the characteristics of the renewable resource while ensuring security of supply. Understanding how to plan and achieve 100% renewable and low emission energy generation is a prerequisite to building sustainable remote communities in the Northern Territory (Figure 14.6).

Understanding how to address immediate targets such as diesel substitution as well as the longer-term ability to secure the Territory's energy future is the basis for this component of the Roadmap.

14.4. "IRAQ DREAM" HOMES

Iraq Dream Homes, LLC (IDH) was established in 2010 to target the enormous potential to develop low-cost and affordable housing projects in Iraq.

IDH is a project company of One Alliance Partners (OAP), which is comprised of several related companies providing an integrated management

Figure 14.7 *Graphic illustration of a "sustainable community" within Iraq. (Iraq Dream Homes, LLC (IDH) 2010).*

approach toward property development, infrastructure development and renewable energy development projects globally. Currently OAP is focused and aligned with specific projects in the Middle East, Southeast Asia and the United States. Iraq Energy Solutions, LLC (IES), another subsidiary of OAP, has launched a renewable energy development initiative to work in concert with Iraq Dream Homes' focus on building "sustainable communities" within Iraq (Figure 14.7) and will provide efficient energy production through privatized distribution.

Figure 14.8 *Solar panels power street lights in Fallujah, Iraq. (U.S Army 2010).*

IDH's managing directors have more than 53 years of professional experience in the fields of project/construction management, architecture, planning/urban design and real estate development. The multidisciplinary background of the company's key staff empowers it to provide excellent services in all aspects of land and real estate development, infrastructure construction, large scale planning, project management and general construction management (Iraq Dream Homes, LLC (IDH) 2010).

14.5. DANISH DISTRIBUTED INTEGRATED ENERGY SYSTEMS FOR COMMUNITIES

The world's energy system is on the verge of a major transformation as a result of climate change and resource scarcity. In order to massively reduce CO_2 emissions, there is a need to build a new energy system that is based on a greatly expanded use of renewable energies. It is almost certain that in 20 or 30 years' time the world will have a very different energy system from the one that currently exists.

In discussions of climate change, it is frequently stated that it is urgent to reduce CO_2 emissions by 40% below what were 1990 levels by the year 2020, and further still to 95% by 2050. Northern countries are undoubtedly the main emitters, which need to implement these reductions.

The technological building blocks for the transition to a sustainable energy future already exist in the form of decentralized cogeneration plants, wind turbines, large and small biogas plants, solar energy and various types of biomass for energy purposes. The primary task, therefore, is to integrate the various forms of renewable energy, sometimes in combination with natural gas, in order to achieve the maximum utilization of renewable energy sources and supplies.

The reality of renewable energy is that it is necessary to combine and integrate technologies since no single renewable energy source can sufficiently stand alone. A comprehensive future conversion to renewable energy requires mobilization of all forms of renewable energy installations, including both large and small plants. It is not enough to base development on technologies which are currently cheapest, as this could lead to a unilateral deployment of large wind turbines in particular.

A persistent global attachment to the dominant fossil-fuel based energy system has limited the development of combined fluctuating solar and wind energies into coherent, autonomous systems. One consequence of this is, for example, that renewable energies, when generated in excess, remain unutilized, or even wasted. Wind turbines in regions with high shares of wind energy, for example, may be periodically shut down when the wind turbines produce too much power. Similarly, when combined heat and power production coincide with excess wind energy, an excess power capacity may occur. These problems will become increasingly frequent as more wind turbines feed power into the grid and more CHP systems are utilized. Electric boilers, however, have proved to be a low-cost solution to capture excess energy, by using excess wind power in fuel-efficient heat and power systems.

A lack of balance between supply and demand of power means that there may periodically be an increasing problem of excess power from the combined supply from wind turbines, solar power and CHP. This problem, however, does not need to exist but is caused by lack of political management and coordination. Appropriate forms of public management and control of supply seems best to solve problems associated with fluctuating and intermittent power production. What is required are political solutions with incentives for the wise use of this so-called surplus power, avoiding selling at very low prices to neighboring countries, the establishment of major new transmission lines and integrated systems to match with supply peaks when winds are strong. The various renewable forms of energy (solar, wind, biomass) can provide an alternative to fossil fuel when they are used in combination with one another. None of the renewable energy forms are capable of covering the need for electricity, heat and transportation if they are used alone. There must be, however, a multiform effort involving many kinds of supply systems, energy storage and saving mechanisms, as well as appropriate user-management strategies.

The constant expansion of renewable energy not only provides realistic potential, but ultimately could lead to an end of the combustion of fossil fuels. The economic risk connected with investments in new conventional power plants will grow significantly, especially with increased uncertainties in the future of fossil fuel prices. Not least will be the growing risk of low utilization, as their annual production is being marginalized by increased use of decentralized renewable energy forms.

Renewable energies will have the key role in the global push toward a carbon dioxide-free future of energy production. Due to their in-principle unlimited potential, in comparison to the current global energy regime, they are treated in this chapter as the primary source of supply for meeting the future demand for electricity, heating and mobility, irrespective of their intermittent character. In areas with high shares of wind or solar availability, these energies will more and more be seen as base load that covers the supply of power by 100% and often more.

Because biomass functions as an ideal long-term storage solution, and due to its limited availability, it is necessary that it be reserved for combustion in combined heating/cooling and power stations with efficiencies of 85% or more. Their primary function is for balancing by up-regulation when solar and wind energy cannot cover the base loads.

Besides grid for electricity, in the future energy structure extensive pipe networks for district heating and cooling will have ancillary functions. Due to their low efficiencies of 40% or less, conventional condensation power stations will not have an important role to play.

Electricity storage will be an essential part of the integrated systems that see power supply, mobility, heating and cooling as a whole, together with existing possibilities like demand-side management. These systems should be affordable, sustainable, and efficient.

Some regions and even countries already have relatively high shares of fluctuating power supply. By 2010, Denmark has seen 22% of its demand for electricity from wind turbines, which by 2015 will grow to around 35%. At low peak power demand and high wind speeds the wind power can currently fully cover the consumption of electricity; at the local level the share of wind power may even be 300% of actual consumption. Interregional compensation with strong connections to neighboring countries still play an important role for up-regulation and down-regulation; it may be a short-term solution, however, as the present importers of excess power most likely in the future will be less interested in buying power as the deployment of fluctuating forms of renewable energy will only increase in neighboring countries as well. The reality is that outlets for periodical overcapacities will be required globally.

Currently many different energy storage systems exist, but only a few are functional and commercially available. Moreover, these technologies need to be compared by their investment volume, their losses and their potential for centralized and decentralized applications. The storage solutions have to be discussed by their limits, environmental effects, geographical requirements, application focus, investment complexity, and efficiency. Furthermore storage technologies have to be optimized in terms of size and capacity, responding time and flexibility, as well as their cost-effectiveness.

The following sections will focus on increased applications of various forms of renewable energy, solutions for power balancing technologies with references to pioneering countries that are already facing the need for new kinds of power management, and its opportunities. Besides storage technologies power production in combination with heating and cooling will be discussed as especially important ancillary solutions. Worldwide, their role is still limited, but with the expected significantly increased use of intermittent and fluctuating energy forms and structural aspects including CHP connected to district heating and cooling they are indispensable.

The supplies of water, electricity, gas, heat and energy for transportation have in common the fact that they are all daily necessities for domestic consumers, as well as for industry and public sector institutions. Therefore, it is the case in many countries that the same company, often a municipal company, may have supplied all these services for several decades, a process which has generally worked to everyone's satisfaction. We will seek to explore these questions very concretely in relation to energy in general and renewable energy in particular.

Denmark is well known internationally for its wind industry and the high share of electricity that is obtained from the wind. District heating together with the production of combined heat and power (CHP), may in the long term prove to be even more important. *This transition represents the single most important initiative to reduce CO_2 emissions in Denmark.* Moreover, the change to combined heat and power supplying up to 60% of electricity and 70% of the demand for heat has created the necessary infrastructure that gradually can be transitioned entirely to renewable energy.

By 2007 43% of the 36 TWh of electricity used in Denmark came from independent power projects (IPPs). Of the 43%, wind power accounted for 18–20% and local CHP around 25%. As a consequence the central power utilities (now owned by Vattenfall and DONG Energy) had their share of the electricity market reduced to a little over 50% of the domestic demand.

It took only 10 years to dramatically shift almost half of the power production from inefficient, centralized, fossil fuel power supply to local, municipal or consumer-owned companies. Coincidently this is the amount of time it takes to build one nuclear power plant, or the equivalent of 1200 MW_{el}. Denmark has not and is not planning to build atomic power plants; this very problematic source of supply was ultimately withdrawn from national energy plans in 1985.

Denmark has succeeded in stabilizing its primary energy supply during 30 years of economic growth. During this time, small CHP plants and renewable energy were introduced and supported by the state. In the period 1975 to 2000 fuel consumption for heating in households was reduced by 30%. In the same period, what was almost a 100% oil-based primary energy supply in the year 1975 decreased the share of oil by means of diversification to 40%. The fuels used were a mixture of oil, coal, natural gas and renewable energy. During the whole period the national gross energy consumption had a level around 20 million tons of oil equivalents (TOE).

The combination of energy conservation and district heating based on CHP were the most important factors in improving overall energy efficiency. Insulation of buildings contributed to a 12% decreased heat demand from 1975 to 2000, while at the same time the heated areas increased by 46%. District heating increased by over 50% while CHP plants replaced boilers, which during the 25-year period, decreased the consumption for heating per m^2 by half. Approximately 40% of this decrease was attributable to the implementation of new building codes that were important for the energy conservation that was as a result of improved energy efficiencies and to new CHP for the balance.

During the 1990s, CHP plants were built in towns and villages as small as 150 households. The small Danish CHP plants are typically built in connection with district heating systems in towns and villages. The CHP plants contain one or more CHP units, peak load boilers and heat storage systems. The CHP units are either engines, gas turbines, or in some cases steam turbines or combined cycle plants.

14.5.1. The Consequences of Fluctuating Power Supply

On days with a high demand for heat combined with high wind, the combined heat and power plants, CHPs, and the wind turbines together sometimes feed more power into the grid than needed by the consumers. The CHPs are on such occasions not operating to cover a need for power but to supply heat to the consumers through the district heating system and the production of electricity can be seen as residual. The wind turbines deliver their electricity production to the grid in accordance with the prevailing wind speeds. CHP and the fluctuating wind and solar together feed their production into the same grid.

The early application of solar and wind to an existing power grid can be balanced without special problems. Once the share of wind energy exceeds 20% or more, however, initiatives have to be taken. With 20% of wind power at the annual basis there will be hours and even days when the wind power production can fully cover the actual need for power. In regions with a high concentration of wind power installations, they can deliver most of the base load for power, but also periodically when wind speeds are high can even produce up to 300% of actual power needs. The structural and technological challenges involved in a transition to renewable energy are obvious when taking into account that in the transition away from fossil fuels and atomic energy, some countries have plans of 50% wind and solar in the supply of power.

Some measures that can be taken to maintain a balance between supply and demand include:

- Store the electricity for use in periods with insufficient solar and wind
- Stop temporarily the operation of some of the wind turbines
- Export power to neighboring countries
- Encouragement of demand-side power consumption
- Find new applications of fluctuating electricity for industrial purposes and in the heating/cooling sector

Special focus will be made in this chapter using fluctuating power in the heating/cooling sector and the infrastructural changes that it requires. Storage of electricity will be discussed separately whereas discontinued operation of windmills is considered nonacceptable during periods when fossil fuels are being combusted. Export of power to neighboring countries involves heavy investments in long-distance transmission lines and is not a long-term realistic solution; periods with high wind speeds is a trans-border phenomenon and neighboring countries are often expanding their wind power capacity as well. With tariff differentiation, industrial consumers and households may be encouraged to change the pattern of use of electricity by operating special machinery, doing the washing at night and charge-discharging future electric cars at periods according to the power supply situation. Experiences, however, indicate that change of consumer behavior has its limits. Electric car owners, for instance, may prefer the benefits of leaving their home in the morning with a fully charged car instead of earning a few cents per kWh peak power delivered to the utility.

Supply and demand on windy days can be balanced with simple solutions such as excess wind and solar power being used in electric boilers and heat pumps at combined heat and power stations. In the case of Denmark, with its many hundreds of CHP stations, it is possible to stop gas engines at periods when the wind and solar power is sufficient for both the actual supply of power, heat and hot water based cooling. As the CHP stations already have big hot water storage tanks, additional excess power can be stored for some days of hot water supply with no electricity to water conversion loss.

As an example, the gas-based cogeneration in Denmark from around 600 decentralized CHP plants and more than 170 industrial auto producers can be stopped and started within minutes so that they match ideally with the

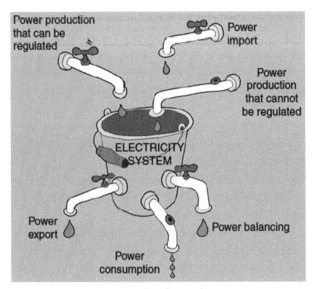

Figure 14.9 *The electrical system.*

fluctuating renewable energy supply. Conversely, conventional fossil fuel, in particular atomic energy power plants, may need several hours or even a day for adjustment. A decentralized power supply system will only work to further create a more robust power structure against national black-outs. *Energinet.dk*, the responsible Danish national power system, plans to divide up the national supply system in cells with each 50–100,000 consumers in primarily self-supplying cells based on local wind and CPH to take advantage of the decentralized and diversified structure.

Figure 14.10 *(left) Thisted combined heat and power for town of 25,000 residents. Fuels used include household waste, straw, wood and geothermal. (center) 6 MW$_{el}$ steam turbine. (right) Trucks discharge waste.*

The challenge for the system operator is, and will continue to be in the future, to further optimize the available resources even more. While Europe as a whole has 11% of the efficient combined heat and power (CHP) in the energy system, the share in Denmark is the world's highest at 45–60%, which at the same time represents not only the problem mentioned but also its solution. Therefore, already existing challenges and solutions as they can be observed in Denmark will throughout the section be presented as examples of reference.

Denmark is the first country to experience problems related to fluctuating renewable energy supplies, and therefore is focused on globally for the quantification of the challenge. In principle, however, the experiences and solutions in the case of Denmark are generally transferrable and will hold true in other countries and regions as well. It should be mentioned that Preben Maegaard is a resident in a municipality of 46,000 inhabitants that since 1992, basically has received its supplies of heating and power from various local renewable forms.

Fourteen (14) central power stations and a small number of windmills covered the production of electricity in Denmark in 1985. By 2009, power production was decentralized with approximately 3,000 windmills and more than 700 CHP units over 0,5 MW_{el}. Together wind power and small CHP can deliver 45% of the annual demand for electricity.

It is often mentioned in the media that with the expansion of wind energy, windy days or periods of a surplus of wind energy will become more and more frequent and contribute to instability in the power system. It is, however, impossible to determine whether the wind or whether the heat-generated power causes the power overflow, as it is to separate the cream from the coffee. By combining and integrating CHPs with electric boilers, heat pumps and hot water storage with the collection of fluctuating energy forms of wind and solar, it will be possible to have much higher shares of solar and wind in the energy structure as a whole, without causing instability in the power system. With further integration of the mobility sector using cars with batteries, additional possibilities of handling fluctuating power will become available.

Integration of fluctuating power production with the heating sector will gradually allow significantly higher future shares of wind and solar energy in the system because in a temperate climate the demand for heat exceeds the need for electricity by a factor of three or so. With increased use of hot water for cooling when applying absorption heat pumps that are driven by hot water, not only does it become realistic for more

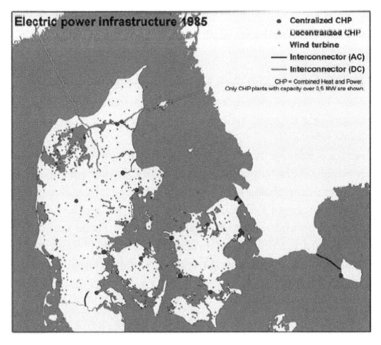

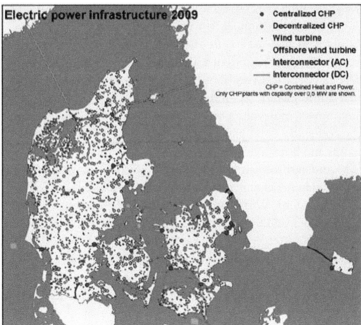

Figure 14.11 *Denmark power stations.*

temperate climate regions, but a more general application is feasible when combining fluctuating power supply from solar and wind with local CHP. Thus the need for heating, cooling and hot water may become the largest single outlet for the disposal of power fluctuations from solar and wind. Furthermore this can be achieved with low initial investments especially when a district heating/cooling network is already available. Building of new district heating structures will at the same time create the needed flexibility for the management of a 100% renewable energy supply and is an affordable, highly efficient and well-developed solution with a mature technology. In regions with sufficient pumped storage capacity this will most certainly be the preferred solution, but it has its limits due to topographical conditions.

With wind and solar as the primary sources of energy, biomass that is easy and cheap to store will be the ideal back-up fuel. On the other side, biomass should not be combusted when wind and solar is sufficient. Environmentally and economically the conversion of excess wind power for the local district heating supply and in their hot water reservoirs will have the per kWh value of the substituted combustible fuel. Thus it becomes an optimal solution instead of exporting the excess power to neighboring countries, sometimes at low or even negative spot market prices.

Following a decision by the EU Commission, use of wind power for heat has become an attractive alternative. Previously, heavy taxation prior to 2007 did not make the excess energy economical. While excess electricity sometimes was sold at the spot market for EUR 0,01 per kWh or less it will at any time (when used for heating) have a value equivalent to the fuel that it replaces, EUR 0,04 to 0,12 per kWh.

Figure 14.12 *Biomass from the agriculture and forests is a limited resource. Dry biomass is ideal for long-term and seasonal storage of energy when solar and wind energy is not available.*

14.5.2. Hot Water Storage

Special focus has been made on *hot water storage* at temperatures below 100°C at atmospheric pressure in insulated tanks made of steel or concrete. Hot water storage within this temperature range is not suitable for electricity production. Considering that the amount of energy needed for space heating in temperate climate is about three times higher than the energy need for electricity, hot water supply and storage based on solar and wind peaks can find large-scale applications by replacing oil, natural gas and coal for heating purposes and for cooling as well by absorption chillers. Several hundred hot water storage tanks in Denmark are installed at local CHP instillations; sizes range from 10 m^3 up to 30.000 m^3. The dimensioning criteria is often to cover the CHP station's need for supply to the district heating network during the power low peak period on weekends. Compared to the earlier mentioned storage solutions, hot water storage involves an especially low initial investment. Compared to batteries and hydrogen the investment may be more than 90% lower. In pumped storage, air compression and flywheels, the conversion losses range between 30–50%, but with initial costs significantly lower than with the hydrogen/fuel cell option.

When the conversion from electricity to hot water is made by electric boiler, the conversion factor is 1:1. When conversion is by compression heat pumps the conversion factor is significantly better and may be 1:2 up to 1:6, depending on the heat source available for the heat pump. A heat pump costs five to eight times more than electric boilers in initial investment;

Figure 14.13 *Viborg CHP plant (left) using natural gas turbine for the production of heat and electricity. The hot water storage is seen to the right of the Turbine-Hall. The city of Odense with its 165,000 inhabitants was a pioneer within large urban CHP resulting in low heating costs. Fuels for the supply of steam to the turbines are straw, household waste and natural gas.*

therefore the annual hours of operation for heat pumps must be 3 – 5,000 hours per year, whereas electric boilers will be profitable with much fewer operational hours per year.

District heating and cooling, CHP and the fluctuating energy forms solar and wind energy, can be combined and integrated to create truly autonomous systems. Pioneering projects demonstrate that a decentralized heat and power system using biomass and wind energy can be cost effective and pave the way for a sustainable energy supply. It has been demonstrated as well that conventional fossil fuel based power production from big centralized power plants can be phased out with improved safety of supply and prevention of climate changes.

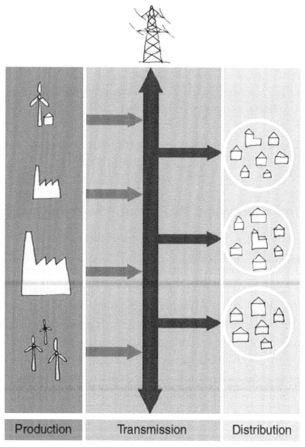

Figure 14.14 *Legislation in 2004 in Denmark resulted in separation of production, of power, transmission and distribution.*

The transition from a few large-scale centralized, fossil fuel based power stations to thousands of wind turbines, solar installations for heat and electricity, combined heat and power stations using biomass or natural gas will be a technological and structural challenge. As was seen in Denmark, the shift to renewables takes some decades, using a mix of conventional and renewable energy technologies. Denmark has by 2010 around 7,000 MW of centralized power plants, 3,000 MW independent combined heat and power and 3,400 MW wind power that will increase to 4,500 MW by 2012. With more and more decentralized CO_2 neutral capacity, the centralized plants can gradually be phased out, which will stabilize consumer energy prices and fulfill international climate commitments.

Production:

- Central power plants owned by DONG Energy (76% owned by Danish state), Vattenfall (owned by Swedish state)
- Municipal and local consumer owned CHP and wind power owned by IPPs

Transmission:

- Power transmission (over 60 kV) is the responsibility of energinet.dk, a new wholly state-owned company

Distribution:

- Distribution is the responsibility of local nonprofit cooperatives, municipalities, or companies with concession

14.5.3. Wind Energy and Its Role in Power Production

Wind energy will be a cornerstone in future supply of electricity. Worldwide, wind energy capacity in 2009 reached 159,000 MW, out of which 38,000 MW new capacity were added. Wind power showed a growth rate of 32%, the highest since 2001. The trend continued that wind capacity doubled every three years. All wind turbines installed by the end of 2009 worldwide are generating 340 TWh per year, equivalent to the total electricity demand of Italy, the seventh largest economy of the world. The wind energy sector in 2009 had a turnover of EUR 50 billion and employed 550,000 persons worldwide. In the year 2012, the wind industry is expected for the first time to offer 1 million jobs. A total wind capacity of 200,000 megawatt was exceeded by the end of 2010. With the accelerated development a global wind power of 2,000,000 MW is possible by 2020.

Contemporary wind technology was born and came of age in Denmark. In the 1980s and 1990s, a dynamic production and implementation process

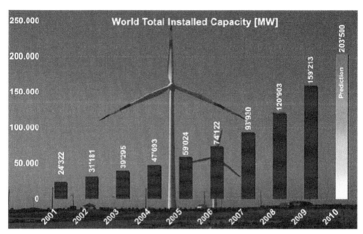

Figure 14.15 *World wind power growth rates and total installed capacity 1998–2009.*

evolved, resulting in the creation of a new industry. By 2010 3,400 MW of wind power was installed. Wind power at days with strong wind and low peak power consumption may cover almost the total need for electricity in Denmark and still normal power quality standards are maintained.

Even if wind power in Denmark was close to a standstill from 2002 to 2009, Denmark by 2009 still was the leading wind power country on a power percentage and per-capita basis. By 2010 there were 3,400 MW installed capacity in Denmark; with a population of 5.5 million, this is equivalent to 618 watts per person. As an extreme, the peninsula of Thy in Denmark with its 46,000 inhabitants, the 225 wind turbines' capacity of 125 MW results in 2,700 watts per person, sufficient for 80% of the demand for electricity coming from the wind. In Germany 2009, there were 25,800 MW and 82 million people, resulting in 314 watts per person. On the other end of the scale, in the UK 4,100 MW were installed, meaning 74 watts per person. Even with the six years of wind energy moratorium Denmark had, by the end of 2009 it still maintained number nine on the world wind top-ten list.

14.5.4. The Wind Energy Development in Denmark

In the meantime new targets had been set for wind energy in Denmark. The wind energy policy by 2008 consisted of two targets to be reached by 2009–10 with the installation of two offshore wind farms, with one of 200 MW in the North Sea, owned by DONG Energy and one of the same size in the Baltic Sea owned by E.On. Together they will increase the

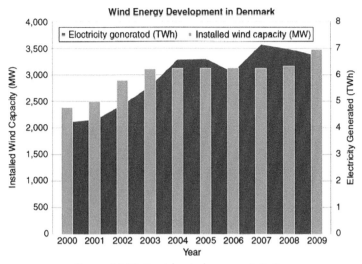

Figure 14.16 *Danish wind energy statistics.*

share of electricity from the wind by 4 % and are supposed to provide power for 350,000-400,000 households. Construction began in 2008 and one of the wind farms was commissioned in 2009, and the second came online in 2010.

Another offshore project of 400 MW in Kattegat has been approved for implementation in 2013, which will increase the share of wind electricity to nearly 30% of the national total. The Danish share of a Swedish-German-Danish joint project in the Baltic Sea, Kriegers Flak, that is currently being negotiated, is expected to be 600 MW. When finished by 2015 another 5–6% of power from wind can be added to a national total, to bring it to around 35%.

In addition to the offshore projects mentioned, onshore installation of wind power will be added. The policy for onshore wind power, disgracefully called "sanitation," foresees the replacement of around 900 medium-size windmills (of up to 450 kW) by 150–200 new megawatt-class turbines. Around 175 MW of wind power capacity will be decommissioned while 350 MW will be installed, resulting in a net increase of 175 MW. With the decided new offshore and onshore capacity, the adaptation of the entire power system to balancing the fluctuating energy forms is more urgent in Denmark than other countries.

Denmark does not have a coherent feed-in tariff scheme. Biogas finds one solution, onshore wind a second that may be quite different from

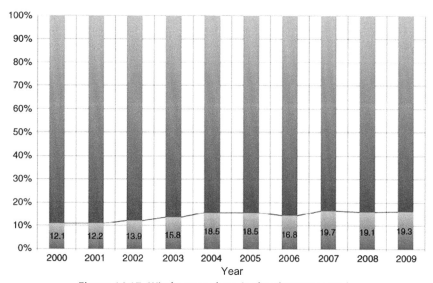

Figure 14.17 *Wind power share in the electricity supply.*

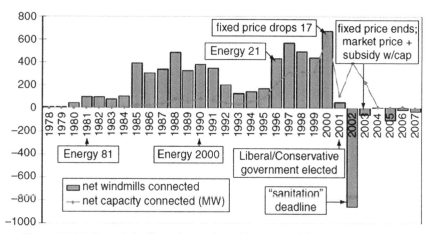

Figure 14.18 *Net windmills and capacity grid-connected by year in Denmark.*

offshore. For PV household, small wind power and mini CHP up to 6 kW installed capacity the net–metering principle is the rule.

The tariff for wind energy in Denmark depends on several variables: on which year the turbine went into operation, how many full-load hours they already delivered, and whether they are offshore or onshore. The tariff comprises a market power price element, power balance compensation and a government subsidy.

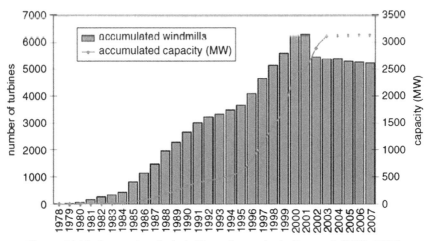

Figure 14.19 *Accumulated windmills and capacity in Denmark (1978–2007).*

New wind turbines onshore receive as a general rule a price premium of EUR cents 3.3/kWh for 22,000 full load hours. Additional EUR cent 0.03/kWh in the entire lifetime of the turbine, to compensate for the cost of balancing etc., leads to a total remuneration of EUR cents 0.7/kWh. In comparison, the most recent offshore Anholt 400 MW wind farm installed by the central utility DONG Energy after a tendering process obtained a guaranteed price of EUR cents 0.14/kwh for 50,000 full load hours, making offshore wind power at least two times more expensive than when placed onshore.

14.5.5. The Ownership Model behind Two Decades of Success

The absence of financial investors early made the wind sector in Denmark unique compared to other countries. At the turn of the century, around 150,000 households were co-owners of a local windmill. The ownership model rather than the technology and tariff schemes were substantial for the success of wind energy in Denmark. It was the key factor behind the high public acceptance that wind power projects enjoyed during that time. It also enabled a much faster deployment, since large numbers of people were involved in the sector that provided the necessary good will.

In 1992 the role of cooperatives shrank due to a change in planning procedures. In 1998, due to liberalization in the sector, the ownership model changed dramatically. The residential criteria for ownership were abolished, and everyone was allowed to own as many windmills as they could get building permission for, anywhere in the country. The take-over bids began, resulting in a dramatic decrease in public involvement. As a result, in 2009 about 50,000 households were co-owners of windmills compared to 150,000 a decade earlier.

Consequently, the attitude toward wind power suffered a reversal. The erection of a windmill often became a local drama and resulted in bitter conflicts that led to long delays or cancellations. In 2008 a wind turbine neighbor compensation scheme was introduced that also caused local conflicts. The solution seemed to be to "normalize" the wind energy sector so that wind power like district heating, CHP and power distribution, changed from investment to public supply, with municipalities and consumer owner local companies as the future wind power owners. This would even make more wind power significantly cheaper.

Denmark already has hundreds of consumer-owned local energy supply companies for combined heat and power, district heating and power distribution. New organizational structures and nonprofit ownership models to the direct benefit of the involved municipalities may prove to be the most realistic long-term solution for community wind power as well to again make it locally acceptable.

14.5.6. CHP and Its General Application

Cogeneration/CHP is the second cornerstone in future self-sustaining supply structures. CHP is the simultaneous production of electricity and/or useful heat and cooling from the same fuel; it more than doubles traditional conversion efficiencies. This chapter focuses especially on two aspects of the combined heat and power (CHP). Its higher overall fuel efficiency than condensation power production and, due to its generally local character with proximity of the plants to the places of consumption, it will have an important role for power balancing in future energy systems with high shares of fluctuating power from wind and solar energy.

As a whole, the European Union generates 11% of its electricity using cogeneration. However, there are large differences in the member states with variations of the share of CHP between 2% and 60%. The four countries with Europe's most intensive cogeneration shares are: Denmark, Latvia, Finland and the Netherlands. The Danish CHP share varies from 43 to 60% with the lower share generated in years when export to neighboring countries of coal-based electricity from conventional power stations is high.

Figure 14.20 *Faaborg CHP plant (left) with its location close to the center of the town supplies 7,000 inhabitants. The circular center building is for the hot water storage (right).*

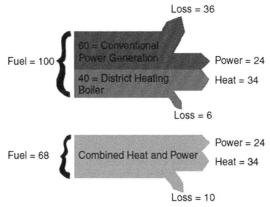

Figure 14.21 *For the same supply of heat and power, with the conventional solution with separated heat and power production the consumption of fuel is 47% higher.*

Other European countries are also making efforts to increase their overall energy efficiency by increased use of CHP. Over 50% of Germany's total electricity demand could be provided through cogeneration. Germany has set the target to double its electricity cogeneration from 12.5% of the country's demand for electricity to 25% by 2020. Due to the advantages of combining the fluctuating energy forms with CHP, Germany's ambitious wind and solar energy programs will not be in conflict with an increased share of CHP.

On the contrary, with an extensive increase of hot water based district heating and cooling and the technologies properly integrated with electric boilers, heat pumps and hot water storage, the CHP stations will pave the way that significantly more solar and wind power can be installed in

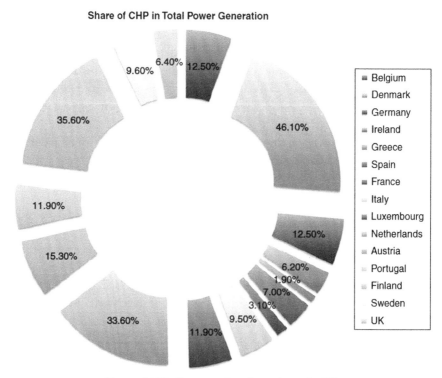

Share of CHP in Total Power Generation

- Belgium
- Denmark
- Germany
- Ireland
- Greece
- Spain
- France
- Italy
- Luxembourg
- Netherlands
- Austria
- Portugal
- Finland
- Sweden
- UK

Figure 14.22 *Cogeneration throughout the EU.*

Germany. With high shares of wind power some of it will, at periods with high wind, be available for the supply of heat. This will lead the way to 100% renewable energy in the largest European economy. Using well-proven already existing technologies, investments per MW new installed capacity of electricity and heating will be in the same size of order as conventional fossil fuel based investments for similar capacities.

Figure 14.23 *Community biogas plant with CHP. The tanker Truck (left) collects 400 tons of liquid waste from farms daily. In the 7.000 m³ Bigadan digester (center) the biogas is produced which the Jenbacher gas engine (right) Converts into heat and electricity.*

Conventional power generation is based on burning fuel to produce steam. It is the pressure of steam which turns the turbines and generates power, in an inherently inefficient process. Because of a basic principle of physics in practice, a maximum of 40–45% of the energy of the original fuel is converted to electricity. The European average is 35% and the world average 30%, meaning that conventional condensation power plants represent an outdated technology.

Cogeneration, in contrast, makes use of the excess heat. With efficiencies up to 90%, CHP is the ideal means of generating heat and electric power at the same time from the same energy source, and it can double or triple conventional power generation.

The favorable environmental implications of cogeneration stem not only from its inherent efficiency, but also from its decentralized character. Because it is impractical to transport heat over long distances, cogeneration equipment must be located physically close to its heat application. A number of environmentally positive consequences flow from this fact: Power tends to be generated close to the power consumer, significantly reducing transmission losses and the need for distribution equipment. Cogeneration plants usually are smaller than conventional power plants, and owned and operated by smaller and more localized consumer owned companies. Cogeneration is at the heart of district heating and cooling systems.

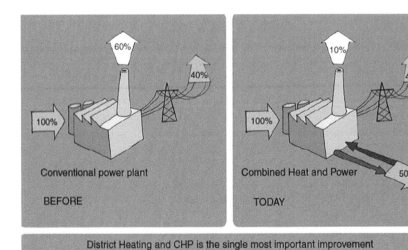

Figure 14.24 *In conventional power plants the loss is 40% or more, whereas CHP may have efficiencies up to 90%.*

Cogeneration units can be manufactured in series like cars, trucks and tractors, which leads to low unit costs, while conventional power stations are unique plants where similar cost reductions cannot be expected.

Due to international policies to reduce emissions of greenhouse gasses, governments, municipalities, owners and operators of industrial and commercial facilities are actively looking for ways to use energy more efficiently. Especially important is cogeneration which is the key technology in future integrated energy systems based on CO_2 neutral energy forms, solar, wind and biomass to maintain continuous supply of electricity, cooling and heating.

Combining a cogeneration plant with an absorption refrigeration system for air conditioning and other needs for cooling allows utilization of seasonal excess heat. The hot water from the cogeneration plant can serve as drive energy for an absorption chiller, and 80% of the thermal output of the cogeneration plant is thereby converted to chilled water. As the hot water replaces electricity for cooling needs, a well-balanced system with low losses can be designed. In this way, the year-round capacity utilization and the overall efficiency of the cogeneration plant can be increased significantly.

14.5.7. Cogeneration Technology

A typical cogeneration system consists of a gas engine, steam turbine, or combustion turbine that drives an electrical generator. Heat exchangers recover the engine's cooling water to produce hot water.

An emerging technology that has cogeneration possibilities is the fuel cell. A fuel cell converts hydrogen or other fuels to electricity without combustion. Heat is also produced. Most fuel cells use natural gas (composed

Figure 14.25 *Ancillary equipment in a CHP station: The exhaust heat exchangers (left) convert the exhaust heat from the gas engines to hot water. The absorption heat pump (right) upgrades 20–40°C water to use in district heating.*

mainly of methane) as the source of hydrogen. Solid oxide fuel cells may be a potential source for cogeneration, due to the high temperature heat generated by their operation and high electrical efficiency around 60%.

14.5.7.1. Cogeneration Units

Cogeneration plants are the modules for combined production of electricity and heat power.

CHP is characterized by:

- Low operational costs
- High efficiency rate, in average 85–90%
- Process automatic control
- 40,000 hours of operation between major overhauls
- Easy installation and operation
- Fulfills international emission standards

Main parts of a cogeneration unit are:

1. Internal combustion engine
2. Engine support frame
3. Alternator
4. Electricity control cabinet
5. Engine cooling system with heat exchangers for connection to the heating system

A CHP module including all components can be factory built on a frame or in a container ready for transport and installation. It can be designed for operating on natural gas, biogas or mixtures of gas. Its compact construction includes all the components of the CHP system, cleaning of exhaust, and can have additional facilities such as heat pumps and an electric boiler to match well with fluctuating energy forms like wind and solar.

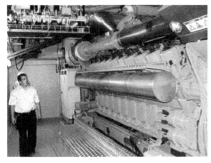

Figure 14.26 *Snedsted CHP (left) with GE Jenbacher V20 3.500 KW_{el} gas motor. Farm CPH (right) with MAN V12 375 kW_{el} motor using biogas.*

A CHP module can be built to industrial standards and meet the requirements of various national energy authorities. The modularized design ensures use of compatible technical components. Trouble-free and rapid commissioning of district heating stations and biogas plants using CHP is possible by using standardized equipment. Building time of conventional power stations may take 4 to 8 years. In the same number of months CHP stations may be planned, installed and commissioned. The CHP module has already been tested and is ready to be connected when it is delivered. Only the lines for electricity and the heat supply must be connected and also to a source of fuel that can be natural gas, biogas and other gases. Consequently, the production of electricity and heat can begin within a short space of time.

14.5.7.2. Cogeneration Applications

Cogeneration systems have been designed and built for many different applications. Large-scale systems can be built on-site as a plant to supply a city or industry with district heating or steam. Cogeneration systems are also available to small-scale users of electricity. Small-scale packaged or "modular" systems are being manufactured for commercial and light industrial applications, and in the agriculture using biogas. Modular cogeneration systems are compact, and can in future be manufactured like automobiles and other mass produced equipment. These systems range in size from 6 kW_{el} and up several MW_{el} capacity. It is usually best to size the systems to meet the hot water and heating needs. They can be operated continuously or only during peak load hours to benefit from peak demand tariffs. Cogeneration is inherently more energy efficient than using separate power and heat generating sources, making it an effective anti-pollution strategy.

Manufacturers of gas engines include GE Jenbacher, Caterpillar, Perkins, MWM, Cummins, Iveco, Rolls-Royce, Wärtsilä, Waukesha, Deutz, MTU, B&W-MAN, Yanmar, Kubota and several other engine builders. Engines can have spark plug ignition which is the most common. Some have 6–10% oil injection for compression ignition. The modern high-speed gas engine has higher efficiency (42%) than gas turbines. The gas engines are intensively used in transcontinental compressor stations with proven long-time trouble-free operation without service and major overhauls. Most of the gas engines are based on a diesel engine platform that is converted: GE Jenbacher is an example of a gas engine designed and dedicated to gas alone.

Especially in Germany, small size CHP using natural gas, biogas, diesel oil, plant oil (PPO), wood pellets or other fuels are used in big numbers. In the city of Freiburg, Germany, often described as Europe's renewable

Figure 14.27 *Two types of small scale CHP in Bavaria, Germany. KW Energy CHP (left) with Kubota 8 KW$_{el}$/18 KW$_{th}$ engine for one family house using plant oil, PPO. Farm CHP (right) with 3x120 KW$_{el}$ (approx.) MAN engines using biogas.*

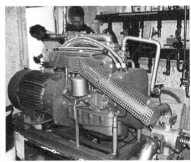

Figure 14.28 *Three types of small size CHP. The Ecopower using natural gas (left), the Sunmachine wood pellet-fueled Stirling CHP (center), and small farm CHP using biogas (right)*

energy capital, a small-CHP campaign for 1,000 CHP units has been launched. They will be installed in Freiburg's old buildings city quarters. Freiburg already has 55% of its electricity coming from CHP consisting of a 40 MW$_{el}$ natural gas CHP of a chemical company and some CHP in public and family houses. The additional potential for CHP in Freiburg is more than 5,000 multi-family buildings and other buildings.

The CHP plants in Denmark already have considerable experience in optimizing their electricity production against the triple tariff which has existed since the early 1990s. Consequently, the plant operators know how to organize the production of the CHP units in order to optimize their income. Meanwhile, Denmark is in the process of replacing such pricing conditions by spot market prices. Consequently, new methodologies and tools for the optimization of the daily operation of small CHP plants are needed. The new markets include up-regulation and down-regulation which meets short-term imbalances mostly caused by increased use of fluctuating and intermittent supply from solar and wind energy.

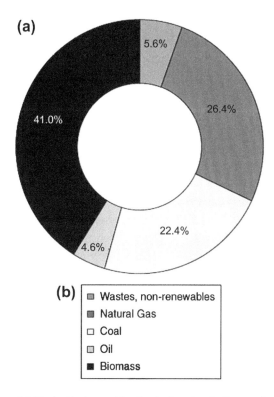

Figure 14.29a.b *Fuels used in district heating in Denmark 2006.*

14.5.7.3. District Heating Networks

District heating means a system supplying heat produced centrally in one or several locations to customers for heating and cooling. It is distributed by means of a distribution network of insulated two-way pipes carrying hot water as a medium. The countries with the strongest growth in CHP capacity also had the strongest growth in district heating capacity. Operators that construct new district heating schemes often consider application of CHP a rational and viable solution or CHP may be a requirement by the local or national energy authorities as it improves energy efficiency significantly and makes it easier to meet CO_2 emission reductions.

The same policies that have created a high penetration of district heating in Scandinavia have also created a high share of CHP with hot water storage connected to the district heating systems. This as well paves the way for increased use of the fluctuating energy from solar and wind leading to 100% renewable energy. Presently power peaks coming from wind and solar can be balanced in the local CPH station as gas engines can respond within seconds or minutes to the variations in the productions of wind and power. In the long term, solar and wind can often cover power base load needs and at periods with maximum supply of solar and wind power, the fluctuating energy forms may even periodically cover both the need for electricity and the need for heating. This saves natural gas or biomass that will be available as back-up fuel when solar and wind is not sufficient.

While some countries have opened up for decentralized energy structures with numerous operators and a diversity of energy drivers, the relatively low penetration of CHP in district heat production, for instance in France, is an example of the obstacles decentralized producers face when entering an electricity market that is dominated by monopolies or oligopolies.

In practices it is feasible for an ambitious national policy to reach, as in Finland, an almost 80% share of district heating produced in combined production. The Finnish example shows that this can be reached by a diversified fuel input without depending on a single fuel type. In the future the integration of large quantities of solar and wind can be included as well, when well planned district heating and CHP offer an important potential for increased energy efficiency in the energy sector.

In Denmark district heat is supplied by some 400 district heating companies, and accounts for approximately 70% of Denmark's heat demand, compared to levels of 30% in 1980. Most of the companies produce and supply the heat, but some purchase heat from the central power plants. The average consumer connection rate in district heating areas is 82% and is still increasing. The district heating network supplies heat to large consumers, apartment blocks, offices, institutions and to single family houses. Danish district heating companies are owned either by the municipalities, particularly in the major cities, or by local consumer co-operatives or foundations.

Forty-three to sixty percent (43–60%) of the electricity generation is from CHP, compared to just under 20% in 1980. Twelve of the 14 central power stations in Denmark deliver all or part of their excess heat to a district heating network. Nearly all large-scale power plants are located close to major cities. This and the fact that 80% of the population lives in urban areas allowed the high shares of combined development of district heating and CHP.

The first steps in the development of CHP were taken in Aarhus, Odense, and Aalborg, the second to fourth largest cities in Denmark. In 1904 the first CHP plant was commissioned, supplying heat and electricity to a hospital. By the mid-1930s, the Copenhagen district heating network was well established, even though heating was to a large extent still provided by small individual coal-fired boilers. The first plant specifically designed for CHP was commissioned in 1934.

The heat planning legislation, initiated in 1979, aimed to increase the share of co-generation in the district heating supply system and to promote use of natural gas. Through the heat plan, the cities were divided into areas suitable for district heating and areas more suited for individual supply of natural gas. The heat plan shielded district heating from inter-fuel competition from natural gas and electric heating.

The majority of district heating loops are owned by the inhabitants of the community. This gives control to the residents and ensures that energy is distributed to the communities at the lowest possible prices. The savings, if profits are accumulated, are given back to the energy consumers in the form of lower heating costs.

Major cities have city-wide district heating systems where almost all of the heat (95 to 98%) is produced in large coal-fired or gas-fired CHP plants and waste incineration plants, with a number of small oil-fired or gas-fired heat-only boilers for peak-load and emergency. Since the early 1980s, no new power plants have been commissioned unless provided with the ability to use CHP and to supply heat to the district heating networks. This was motivated by security of supply and environmental concerns encouraging energy efficiency. Construction of new electricity-generating capacity must be justified by the need for new heat production capacity.

In addition to the large-scale CHP and district heating units, a large number of small-scale CHP plants exist. Most of the plants range between capacities of 0.5 to 5 MW and supply heat to small communities and institutional buildings. Often the plants consist of more than one CHP gas engine unit. Small-scale CHP plants, which are not connected to a district heating network, rarely exceed an electric-capacity of 1 MW. Small-scale CHP plants are laid out to cover at least 90% of the local heat demand. The electricity generated is sold to the public grid. The national power system responsible will purchase the electricity from all types of producers. The main fuels used in small-scale CHP are natural gas and, to a lesser extent, biogas and other biomass. In connection with the presentation of the *Energy 2000* plan in 1990, a more ambitious program for small-scale CHP was put forward. To accelerate the establishment of small-scale CHP, a state subsidy was introduced in 1992 for power production from waste incineration, natural gas and renewables. The subsidy originally amounted to 1.4 EUR cent per kWh. With an implementation period of less than 10 years the development of small-scale CHP peaked in the late 1990s. The capacity was sufficient to produce 25% of the national demand for power.

About 80% of the installed capacity is based on natural gas engines. Sixteen percent (16%) are gas turbines and 4% are biogas-fired engines. Most of the installed gas engines have an electric capacity in the range of 0.5 to 4 MW_{el} whereas gas turbine units typically range in capacity from 5 to 25 MW_{el}.

The electricity from CHPs is fed into the national grid. The feed-in tariffs for CHP were based on a three-tier tariff system, with tariffs reflecting electricity demand patterns (low, medium and peak tariff periods). After liberalization of the electricity sector, market principles were applied as in other Nordic countries.

Industrial CHP is used in industries with high demand for process heat, especially the petrochemical, wood, malt and paper industries. The food industry and greenhouses can also use low-pressure steam or hot water from CHPs. To illustrate the variety of application, the national railway company gets hot water for cleaning its passenger trains from a medium CHP at the maintenance facility. In 2000, electricity production from industrial CHP was about 8% of total power generation.

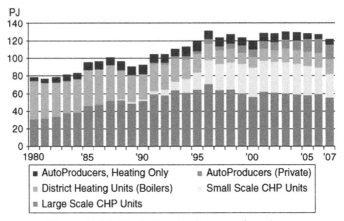

Figure 14.30 *District heating by type of producer.*

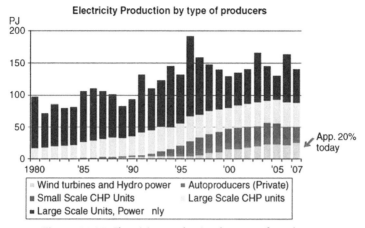

Figure 14.31 *Electricity production by type of producers.*

14.5.7.4. Technology for Decentralized District Heating

As mentioned earlier the preferred CHP technology has predominantly been gas-powered engines. Stationary natural gas engines used in combined heat and power applications boast a factor four reductions in CO_2 emission compared with conventionally generated thermal coal power for the same produced power and individual supply of heat.

This is because:

1. The heat can be used if the system is placed in the community increasing the total efficiency of the system to over 85% compared with the best thermal coal power plant at around 45% electrical efficiency.
2. Natural gas has a tenth of the SO_x, half of the NO_x and a third of the CO_2 produced from combustion of coal. The cost of removal of these pollutants in a coal generation plant is significant.
3. The cost to install a gas engine is around 40% lower per kW_{el} than that of a coal plant and even cheaper if removal of the emissions from coal is included. Additionally, gas fired engines can be installed in 6 months as opposed to 4–8 years for a thermal coal plant.
4. Gas engines are manufactured in big numbers and cheap while central power plants are one-of-a-kind technology. The gas engines can be installed in existing or new buildings without any noise or visual impact for the neighborhood.

With these benefits it became possible for local district heating companies owned by municipalities or consumers to build their own CHPs and offer cheaper heat to households. This became the primary driving force that encouraged a rapid change to local CHP.

CHP using natural gas is not renewable but local CHP creates the basis for a decentralized energy structure that later, with modest investments, can be changed to local renewable energy sources. Stationary gas engines can run on a variety of fuels which can be tailored to local fuel availability.

Local supply sources can be:
- Biodiesel
- Plant oil
- Biogas
- Gasified biomass
- Land fill gas
- Solar thermal installations
- Wind heat and power, WHP

If these alternate fuel sources are not available, natural gas can be used as a transitional fuel while the community determines which fuel can be utilized in the future. In essence, district heating with CHP provides the initial framework toward a renewable energy power and heat system.

14.5.7.5. Summarizing the Advantages of Local CHP

Advantages of community based CHP units are significant, the main benefits being:

1. *Reliability:* Gas engines can be used where reliability is of the utmost importance. Typically these gas engines are installed in transcontinental gas compressor stations, drilling rigs, offshore oil platforms, and villages not served by the national power grid.

2. *Community autonomy:* Having local CHP provides the local community with autonomy giving the "power to the people." This enables the community to ensure that the power is developed in an appropriate manner.

3. *The ability to incorporate renewable energy in the future:* Having CHP with district heating opens opportunities to incorporate large fractions of renewable energy in the form of biogas, solar thermal heating, wind for heat, biomass gasification, plant oil-based fuels and combustion of locally grown biomass.

4. *Scalability and flexibility:* Local CHPs are scalable and flexible to operate. This makes it easy to increase capacity in the future and matches well with the incorporation of wind and solar power in the supply system.

5. *High efficiency:* Medium size stationary gas CHP units boast an electrical efficiency of around 42% and with heat recovery of the jacket water,

exhaust, lube oil and turbo charger can achieve an overall efficiency of over 85% (power 42% plus heat 43%).

6. *Cost-effective heat and power:* With high total efficiencies and two energy products from the same fuel source, cost of power and heat can be reduced. As an example in 2007, Denmark according to Eurostat had the fourth lowest power prices (without taxes) for GWh-consumers in Europe with Sweden, Norway, France and Finland being lower.

14.5.7.6. Combining CHP and Wind

On days with a high demand for heat combined with high wind, the CHPs and the wind turbines, as mentioned earlier, together feed more power periodically into the grid than needed by the consumers. The challenge for the system operator in some countries is to optimize the available resources. While Europe as a whole has 11% of the efficient combined heat and power (CHP) in the energy system, the share in Denmark is 43–60%, the world's highest, which at the same time represents not only the problem mentioned but also its solution.

The combined high CHP and wind power production causes a potentially major problem in the power sector. However, supply and demand can be balanced by feeding on windy days the excess wind power into electric boilers and heat pumps at the combined heat and power stations which makes 100% renewable energy supply of heat and power realistic. The electric boilers and heat pumps make it possible to stop the natural gas and biomass fueled CHP units at periods with excess wind and solar energy. With increased renewable energy shares, in the future more and more often

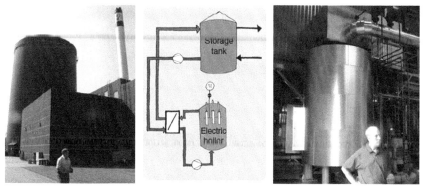

Figure 14.32 *Principle of down-regulation using excess wind power at a CHP plant with 10,000 consumers. Storage for 2 days heat supply (left) in Grindsted; 20 MW electric boiler (right).*

the fluctuating solar and wind energy will be sufficient to satisfy both the need for power and heat.

Compared to other uses and storage of excess power, electric boilers and heat pumps cause no conversion loss. Heat pumps even boost the excess production of power by a factor of 2 to 5 depending on the energy source available. For heat pumps, a low quality source may be atmospheric air, soil, and water from the sea or rivers which will result in a relatively low efficiency while geothermal heat, air ventilated from buildings and industries and other types of potential heat sources up to 40°C will result in high efficiencies.

As they require substantial investments, the operation of heat pumps using excess wind and solar power for balancing the grid by down-regulation should result in yearly operation of 4,000 hours or more. This amount of excess power will in a more distant future be available at medium peak periods. As the fluctuating solar and wind will always cause short term high peaks, the electric boilers will, due to their low initial investment, be the preferred solution to take the high peaks coming from solar and wind energy.

Therefore it is not a relevant discussion whether electric boiler or heat pumps should be preferred. In the early stage of using fluctuating solar and wind in the heating system, electric boilers will be optimal, while increased amounts of fluctuating power will pave the way for heat pumps to convert the medium peak electricity into district heating while the electric boilers will take the high peaks.

Like the transition from fossil fuels to renewable energy forms leads to zero emission, the modern power balancing is an integrated part of the change of the energy system. The natural gas based cogeneration in district heating stations and the industrial auto producers can be adjusted within

Figure 14.33 *Three sizes and types of electric boilers for power down-regulation: 20 MW, 10 kV (left); 3x300 kW, 400 V (center); 45 kW, 400 V (right).*

seconds or minutes and match ideally the fluctuating renewable energy supply.

New taxation rules can make it economically beneficial to use fluctuating wind electricity for district heating according to a decision by the EU Commission. While excess electricity sometimes is being sold at the spot market for EUR 0.01/kWh or even at negative prices it will at any time, when used for heating, have a value equivalent to the fuel that it replaces or EUR 0.04 to 0.12/kWh depending on the type of fuel type and taxation.

In the Thy peninsula, with its 46,000 inhabitants, the 225 windmills and other renewables cover 100% of the annual need for electricity and the local energy production has become an important source of income. On days with strong wind, the wind turbines may even produce four times more than the actual consumption and the power quality still lives up to the highest standards. The local utility, *Thy-Mors Energi*, has demonstrated real-time management of such big quantities of wind energy to visitors from all parts of the world.

In the towns and villages in Thy, people get their heating from hot water pipelines in the streets. It is environmentally and economically the best solution to use the excess wind power for the actual supply and storage in the big hot water reservoirs of the local district heating suppliers instead of exporting the surplus power to neighboring countries sometimes at low spot market prices.

A New Municipal Energy Foundation plans to own future windmills, and 80 MW are in the planning process. The income of the foundation may be up to EUR 7 million/year earmarked for local energy initiatives which secures acceptance of the wind power and illustrates the benefits of change from investor policy to local supply.

BOX 14.1 Thisted Municipality Renewable Energy Data (2008)

Data Thisted Municipality (2008)

- 225 windmills
- 124.600 kW installed wind power capacity
- 35.800 kW_{el} installed CHP capacity
- Power from wind energy 265 GWh
- Power consumption of 340 GWh
- 80% from wind
- 20% from biogas and CHP waste
- A small amount PV

For the last 30 years, farmers, industry, utilities and cooperatives in Thy have invested in and used renewable energy resources:

- Biomass for district heating
- Biogas, small and large CHP
- Geothermal heat
- CHP waste incineration
- Wind power
- Wind energy management
- CHP and Wind Heat & Power, WHP (planned)

Integrating all the local energy plants in the Thisted municipality in one autonomous system has the highest priority. The further development of wind power will make it the primary source of electricity and heating, WHP. With 80 MW of new additional wind power, peaks will be more frequent. The strong crucial local support is obtained by local ownership of windmills and biogas plants by several farmers. All the CHP plants and the district heating are not-for-profit consumer owned. Local surplus of wind electricity will be used in the future in combined heating and power plants and periodically replace natural gas and biomass. Biomass will function as back-up storage when wind and solar energy is not sufficient to cover the need for electricity and heating as solid and liquid biomass is easy and cheap to store. This will only be used when solar and wind is not sufficient to cover the needs for heat and power.

In the peninsula of Thy, the local community owns/installs/operates the green power producing infrastructure; therefore the benefits from the infrastructure are reaped by the community. Local pollution and CO_2 reduction, job creation, business development, economic diversification and skill building are of general interest for the community. Once a community has experience with community power, that skill can be transferred to other communities (Maegaard 2011).

14.6. RENEWABLES IN AFRICA

For African countries, economic and social development cannot be achieved in the absence of adequate energy supplies. While reliable energy services, by themselves, are not sufficient to eradicate extreme poverty, they are necessary for creating the conditions for economic growth and improving social equality. Access to modern energy services can help reduce spending; because of the inefficiency of traditional energy forms, the poor often pay higher unit costs for energy than do the rich. In addition, women in most parts of Africa spend large amounts of time fetching wood and water, thereby losing time that they could otherwise devote to schooling and revenue-generating activities. Furthermore, indoor air pollution from low quality solid cooking fuels (wood, charcoal, dung, and waste) imposes a major health hazard.

In view of their modular nature and availability at the local level, renewable energy technologies can contribute to sustainable development by increasing access to modern energy services to the majority in rural areas.

Energy access and quality in Sub Saharan Africa (SSA) is relatively low with high costs:

- 560m people do not have access to electricity = 74% compared to 28% in other developing countries
- 625m people do not have access to modern fuels = 83% compared to 33% in other developing countries
- 2m deaths per annum (globally) caused by indoor air pollution (fuel burning for cooking & heat) and the burden of disease in Africa is particularly high
- In 2009, South Africa produced 70% of all SSA's electricity
- Subsidies have failed to bring energy prices down and are not sustainable

There are many renewable energy projects in SSA, as shown in Figure 14.34.

14.6.1. Hydropower

Only about 5% of Africa's hydropower potential of just over 1750 TWh has been exploited. The total hydropower potential for Africa is equivalent to the total electricity consumed in France, Germany, United Kingdom and Italy put together. The Inga River in the Democratic Republic of Congo (DRC) holds great potential for hydropower generation in Africa with an estimated potential of around 40,000 MW. In fact, the DRC alone accounts for over 50% of Africa's hydropower potential while other countries with significant hydropower potential include Angola, Cameroon, Egypt, Ethiopia, Gabon, Madagascar, Mozambique, Niger and Zambia. Despite the low percentage use, large-scale hydropower so far provides over 50% of total power supply for 23 countries in Africa.

Small hydropower systems (SHPs, less than 10MW) can supply energy to remote communities and catalyze development in such communities. In

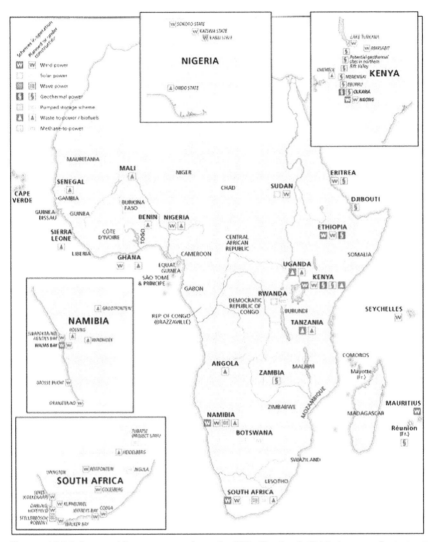

Figure 14.34 *Renewable energy projects in SSA. (Petrie, B 2011, Loccum Protestant Academy, Germany).* (See color plate 46.)

comparison to large hydropower systems, the smaller systems require significantly lower capital costs. This allows increased local private sector participation and community involvement. Most African countries have a large potential for small hydro systems and some are already exploiting it with a special focus on rural communities.

14.6.2. Biomass

Several agro-based industries in the continent, such as wood-based industries, palm and rice mills, sugar, and paper and pulp, use their waste to produce both process heat and power, which in most cases is used locally. Co-generation from agriculture waste holds great potential for Africa. Cogeneration contributes as much as 40% of the total electricity generated in Mauritius.

Dissemination of biogas digesters for household applications has not been very successful due to high capital costs, insufficient feedstock and water, high labor demand and negative public perception, among other reasons. Countries like Ghana, Kenya, Niger, Burkina Faso, Mali, Ethiopia, Senegal and Rwanda have implemented pilot projects aimed at establishing the technical and socio-economic viability of biogas technology as an alternative source of energy for cooking and decentralized rural electrification.

In the case of Ghana, the Appolonia project installed a system that generates 12.5kW electric power, which is fed into a local grid, supplying electricity at 230V for domestic use to 21 houses, street lighting, and five social centers in the community. The biogas is produced from cow dung and human excreta. Two diesel engines of 8kW each were modified to operate on a dual fuel (mixture of biogas and diesel). Project results show that the diesel-biogas system saves 66% in diesel consumption compared to pure diesel generation. The dissemination of biogas digesters to institutions is quite promising, with the private sector already leading the dissemination of biogas digesters in countries like Ghana, Rwanda and Tanzania.

The majority of sub-Saharan households rely primarily on wood fuel for cooking and heating. Wood fuel is the main source of fuel in rural areas while charcoal is commonly used in the poorer urban households. However, shortages of alternative energy sources including electricity blackouts and brownouts often force even the better-off households to use charcoal. As a response to fuel wood shortages, improved biomass cook stoves have been promoted throughout Africa.

14.6.3. Geothermal

Geothermal energy is an untapped renewable energy source that is abundantly present in many parts of Africa. It has a potential of generating up to 14,000 MW from geothermal sources. However, only a few countries such as Kenya have used it commercially. As of today, Kenya has installed up to 127 MW, amounting to about 17% of the national power supply, followed by Ethiopia with a 7 MW installation. Plans to use potential of geothermal energy in Uganda, Tanzania and Eritrea are at different stages.

14.6.4. Wind Power

In terms of installed capacity at the beginning of 2008, Africa only had about 476MW of installed wind energy generation capacity compared to a global estimate of 93,900MW. Countries developing large-scale wind energy projects so far include Morocco, Egypt, Tunisia, South Africa, and Ethiopia.

14.6.5. Solar Power

Large-scale solar energy projects are very limited in Africa because of cost constraints (Figure 14.35). Detailed feasibility studies have established that Africa has great potential for concentrated solar thermal power generation from desert areas like the Sahara, Namib, etc., with competitive power production costs around 4–6c/kWh. So far, only South Africa operates a solar thermal power system plant, generating 0.5 MW. Egypt plans to install solar thermal plants of 30 MW by 2010 and 300 MW by 2020. Several countries in Northern Africa are planning to develop solar thermal plants of varying capacities buoyed by interest from European countries.

Figure 14.35 *Solar charging unit for mobile phones in Africa. (Rolf-Peter Owsianowski 2011, Loccum Protestant Academy, Germany).*

Figure 14.36 *Solar water pumping and lighting station in Mali. (Togola 2011).*

Small-scale energy systems are of two types: The first category produces electricity based on photovoltaics (PV) (Figure 14.36) and wind power, for instance, while the second category produces thermal energy for heating, drying and cooking.

Solar home systems in the household sector are by far the most common application of this kind of system. South Africa and Kenya have some of the highest documented installed capacities of solar PV systems that stand at over 11,000 and 3,600 kWp respectively. Unfortunately, poor households have not benefited as much as high income households from solar PV systems because of the high upfront costs. Some solar thermal systems have been disseminated for water heating and solar cookers. Presently, solar water heaters are predominantly used in Eastern, Southern and Northern Africa for household application and in the hospitality industry.

14.6.6. Biofuels

Various studies have estimated that the potential for sustainable biofuels production (production that preserves biodiversity, rainforests and water resources, and does not endanger food security) in 2050 for Sub-Saharan Africa ranges from 41 to 410 exajoules (Smeets, Faaj, & Lewandowski 2006). The lower range of this estimate is more than twice the total amount of energy that was consumed in Africa in 2008, which is about 19 exajoules. About 39 countries in Africa are net oil importers. Therefore, the development of biofuels will reduce dependency on imported fuels. Agriculture is the mainstay of most countries in Africa, so the development of biofuels in Africa, especially in rural areas that have the land to use, could bring in many potential benefits including increased access to electricity, transformation of the rural economy due to the availability of reliable energy supplies, and new employment opportunities.

Zimbabwe, Kenya and Malawi have ethanol programs in which ethanol produced as a by-product from sugar industries has been blended with petrol and used as transport fuel. Zimbabwe is the only one of the three countries, however, to mandate that ethanol be blended with all gasoline sold and produced up to 40 million liters of ethanol per year. The Kenya plant was closed in 1992 due to lack of viability. At its peak, it produced up to 45,000 liters and some rural employment was achieved.

Other recent efforts to develop large-scale biofuels include palm oil and cassava-based ethanol in West Africa, Jatropha-based biodiesel in Mali, Tanzania, etc. Small-scale biofuels production in Africa is so far primarily based on Jatropha biodiesel. Most countries in Africa have ongoing programs to grow Jatropha in rural areas to produce biodiesel.

14.6.7. Energy Efficiency

Countries in Africa have considerable scope for increasing energy efficiency on the energy supply side and reducing energy consumption on the demand side without decreasing economic output, lowering the standards of living, or diminishing the quantity and quality of social services provided. Studies by the International Energy Agency show that in Africa energy intensity, i.e., total energy consumed per GDP, is at least twice the global average. However, energy efficiency continues to be a peripheral issue in the overall energy sector planning and development in Africa.

14.7 RENEWABLES IN INDIA

Photovoltaic electricity is much cheaper than diesel generators.

Almost 25% of the population in India, according to a report by the International Energy Agency (IEA), is still not connected to the electricity grid, and connected land areas are often affected by power outages. To compensate, individuals and businesses often have diesel generators (Figure 14.37). Solar power is already asserting itself here as a better alternative; thanks to the price landslide, a price of 8.78 rupees (13.8 cents) per kilowatt hour is now well below that of diesel generators, which hold at 17 rupees (26.7 cents) (Marshall 2012).

A key factor for the boom, however, was mainly the promotion of grid-connected solar power projects. The increase in India last year from 600

Figure 14.37 *Photovoltaic system: Options in developing countries. (Photo: FLickr/ Pwrdf).*

million to 4.2 billion dollars almost reached the level of the 4.6 billion dollar wind power development. India's authorities say, 180 MW were connected to the grid, and BNEF expects that up to 375 megawatts will be annexed at the end of 2011. India's long-term strategy also shows the "solar mission" of the government, which wants to install the capacity of 20 gigawatts (GW) of solar power by 2022.

Clean energy investments in India reached $10.3bn in 2011, some 52% higher than the $6.8bn invested in 2010. This was the highest growth figure of any significant economy in the world. There is plenty of room for further expansion; in 2011, India accounted for 4% of global investment in clean energy. The large growth was driven by a seven-fold increase in funding for grid-connected solar projects: from $0.6bn in 2010 to $4.2bn in 2011. Asset financing for utility-scale projects remains the main type of clean energy investment in India, with $9.5bn in 2011. This is significant as the higher lending rates observed over the past year could have negatively impacted asset finance. Venture capital and private equity investment also made a strong comeback with $425m invested in 2011, more than four times the 2010 figure. Wind and solar project developers such as Mytrah Energy India and Kiran Energy Solar Power succeeded in doing deals. The only major type of investment that fell in 2011 was equity-raising via the public markets. Only $201m was raised compared to a record $735m in 2010 when the Indian stock market was at its all-time high.

The wind sector added a record 2,827MW of capacity in 2011 compared with 2,140MW in 2010. This kept India at the third rank globally in terms of new installations, behind China and the USA. Bloomberg New Energy Finance estimates that 2,500MW to 3,200MW of wind capacity could be added in 2012. A substantial increase in grid-connected solar capacity was also observed, up from 18MW in 2010 to an estimated 277MW by end of 2011. In 2012, another 500-750MW of solar projects could be added. India's 11th five-year plan, running from April 2007 to March 2012, targeted the addition of 12.4GW of grid-connected renewable energy. According to Bloomberg New Energy Finance, this target will be exceeded, with 14.2GW capacity installed during the period.

India's record performance in 2011, and the momentum it is carrying into 2012, is one of the bright spots in the clean energy firmament. With support mechanisms falling away in the US, the ongoing financial crisis in Europe, and China already going at full speed, it is gratifying to see some of the world's other major potential markets coming alive. India is firmly in the lead group and we are seeing interest around the world in being part of what is unfolding there (Bloomberg 2012).

14.8. DISTRIBUTED RENEWABLE ENERGY AND SOLAR OASES FOR DESERTS AND ARID REGIONS: DESERTEC CONCEPT

The DESERTEC concept of the pioneer Gerhard Knies aims at promoting the generation of electricity in Northern Africa, the Middle East and Europa using solar power plants, wind parks and the transmission of this electricity to consumption centers, promoted by the non-profit DESERTEC Foundation. Despite its name, DESERTEC's proposal would see most of the power plants located outside of the Sahara Desert itself, in the more accessible southern and northern steppes and woodlands, as well as the relatively moist Atlantic coastal desert (Figure 14.38). The original and first region for the assessment and application of this concept is the EU-MENA region (Europe, Middle East, and Northern Africa). The realization of the DESERTEC concept in this region is pursued by the industrial initiative Dii.

Under the DESERTEC proposal, concentrating solar power systems, photovoltaic systems and wind parks would be spread over the desert regions in Northern Africa like the Sahara Desert. Produced electricity would be transmitted to European and African countries by a super grid of

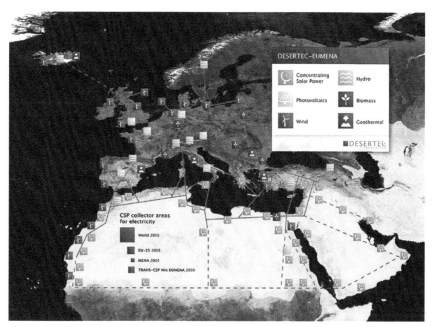

Figure 14.38 *Possible infrastructure for a sustainable supply of power to Europe, the Middle East and North Africa (EU-MENA) (Euro-Supergrid with a EU-MENA-Connection Proposed by TREC).* (see color plate 43).

high–voltage direct current cables. It would provide a considerable part of the electricity demand of the MENA countries and furthermore provide continental Europe with 15% of its electricity needs. By 2050, investments into solar plants and transmission lines would total €400 billion.

The project of the realization of the DESERTEC concept in EU-MENA is developed by Dii GmbH, a consortium of European and Algerian companies founded in Munich and led by Munich Re., and is incorporated under German law. The consortium consists of the DESERTEC Foundation, Munich Re., Deutsche Bank, Siemens, ABB, E.ON, RWE, Abengoa Solar, Cevital, HSH Nordbank, M & W Zander Holding, MAN Solar Millennium, and Schott Solar. Press investigations point to a number of more interested parties, among them ENEL, Électricité de France, Red Eléctrica de España and companies from Morocco, Tunisia and Egypt (Wikipedia 2012).

14.8.1. Scientific Background of the Concept

Main scientific data has been delivered by an international network of scientists, experts and politicians from the field of renewable energies coordinated by Dr Franz Trieb, German Aerospace Centre (DLR), conducted in three research studies to which the author has also contributed.

14.8.1.1. Concentrating Solar Power for the Mediterranean Region (MED-CSP)

To keep global warming in a tolerable frame, the Scientific Council of the German Government for Global Environmental Change (WBGU) recommends in its latest study based on a scenario of the IPCC (Intergovernmental Panel for Climate Change) to reduce CO_2 emissions on a global level by 30 % until 2050.

The advice of WBGU was to establish model projects to introduce renewable energies on a large scale as a strategic lever for a global change in energy policies. A strategic partnership between the European Union (**EU**), the Middle East (**ME**) and North Africa (**NA**) is a key element of such a policy for the benefit of both sides: MENA has vast resources of solar energy for its economic growth and as a valuable export product, while the EU can provide technologies and finances to activate those potentials and to cope with its national and international responsibility for climate protection, as documented in the Johannesburg agreement to considerably increase the global renewable energy share as a priority goal.

A special interest lies in the electricity sector which is responsible for a considerable share of greenhouse gas emissions. A further field of interest is the increasing demand for technically desalted water, which will require increasing energy input to the water supply sector. In order to establish appropriate instruments and strategies for the market introduction of renewables in the European and MENA countries, well founded information on demand and resources, technologies and applications is essential. It must further be investigated whether the expansion of renewable energies would imply unbearable economic constraints on the national economies of the MENA region.

The MED-CSP study focuses on the electricity and water supply of the regions and countries illustrated in Figure 14.39 including Southern Europe (Portugal, Spain, Italy, Greece, Cyprus, Malta), North Africa (Morocco, Algeria, Tunisia, Libya, Egypt), Western Asia (Turkey, Iran, Iraq, Jordan, Israel, Lebanon, Syria) and the Arabian Peninsula (Saudi Arabia, Yemen, Oman, United Arab Emirates, Kuwait, Qatar, Bahrain).

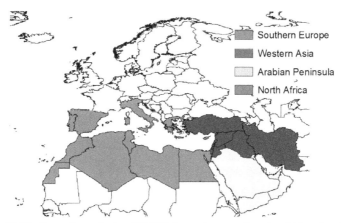

Figure 14.39 *Countries of the EU-MENA region analyzed within the MED-CSP study.* (See color plate 44.)

The results of the MED-CSP study can be summarized in the following statements:

- Environmental, economic and social **sustainability in the energy sector can only be achieved with renewable energies.** Present measures are insufficient to achieve that goal.
- **A well balanced mix** of renewable energy technologies can displace conventional peak-, intermediate and base load electricity and thus **prolongs the global availability of fossil fuels for future generations** in an environmentally compatible way.
- Renewable energy **resources are plentiful** and can cope with the growing demand of the EU-MENA region. The available resources are so vast that an additional supply of renewable energy to Central and Northern Europe is feasible.
- **Renewable energies are the least-cost option** for energy and water security in EU-MENA.
- Renewable energies are the **key for socioeconomic development and for sustainable wealth** in MENA, as they address both environmental and economic needs in a compatible way.
- Renewable energies and energy efficiency are the main pillars of **environmental compatibility**. They need initial public start-up investments but no long-term subsidies like fossil or nuclear energies.
- An adequate set of **policy instruments must be established** immediately to accelerate renewable energy deployment in the EU and MENA (Trieb).

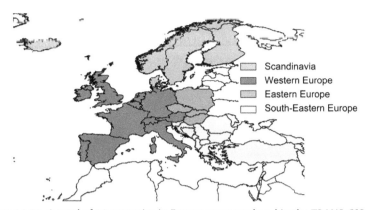

Figure 14.40 *A total of 30 countries in Europe were analyzed in the TRANS-CSP study.*

14.8.1.2. Trans-Mediterranean Interconnection for Concentrating Solar (Trans-CSP) Power (TRANS-CSP)

Competitiveness, compatibility with society and the environment, security of supply and international cooperation are considered main pillars for sustainability in the energy sector. For each of the 30 countries[1] shown in Figure 14.40, electricity scenarios from the year 2000 to 2050 were developed that show a consistent transition to a sustainable supply that is inexpensive, compatible with the environment and based on diversified, secure resources.

Sustainable power in Europe (EU) can be based to a great extent on renewable generation including solar electricity imports from the Middle East and North Africa (MENA). A well-balanced mix of renewable energy sources with fossil fuel backup can provide affordable power capacity on demand.

The transfer of solar electricity from MENA to Europe can initiate an understanding of a common EUMENA region, starting with a partnership and free trade area for renewable energy, and culminating in what H.R.H. Prince El Hassan Bin Talal, President of the Club of Rome, called a "Community for Energy, Water and Climate Security" at the World Energy Dialogue at the Hannover Technology Fair in April 2006.

---

[1] TRANS-CSP covers: Iceland, Norway, Sweden, Finland, Denmark, Ireland, United Kingdom, Portugal, Spain, France, Belgium, Netherlands, Luxembourg, Germany, Austria, Switzerland, Italy, Poland, Czech Republic, Hungary, Slovakia, Slovenia, Croatia, Bosnia-Herzegovina, Serbia-Montenegro, Macedonia, Greece, Romania, Bulgaria, Turkey, www.dlr.de/tt/trans-csp. The MENA countries and some Southern European countries were analyzed within the MED-CSP study, www.dlr.de/tt/med-csp, quantifying the opportunities of the Mediterranean region.

The TRANS-CSP study comprises a comprehensive data base on the present and expected demand for electricity and firm power capacity, quantifies the available renewable energy resources and their applicability for power, provides scenarios of the electricity supply system until 2050, and evaluates the resulting socioeconomic and environmental impacts for each of the analyzed countries. The executive summary at hand gives the aggregated results for Europe as a whole, while individual country data is given in the annex of the full study report.

The TRANS-CSP study analyzes the renewable electricity potentials in Europe and their capability to provide firm power capacity on demand. The concept includes an interconnection of the electricity grids of Europe, the Middle East and North Africa (EUMENA) and evaluates the potential and benefits of solar power imports from the South. The conventional electricity grid is not capable of transferring large amounts of electricity over long distances. Therefore, a combination of the conventional Alternate Current (AC) grid and High Voltage Direct Current (HVDC) transmission technology will be used in a Trans-European electricity scheme based mainly on renewable energy sources with fossil fuel backup. The results of the TRANS-CSP study can be summarized in the following statements:

- A well-balanced mix of renewable energy sources backed by fossil fuels can provide sustainable, competitive and secure electricity for Europe. For the total region, our scenario starts with a reported share of 20% renewable electricity in the year 2000 and reaches 80% in 2050. An efficient backup infrastructure will be necessary to complement the renewable electricity mix, providing firm capacity on demand by quickly reacting, natural gas fired peaking plants, and by an efficient grid infrastructure to distribute renewable electricity from the best centers of production to the main centers of demand.

- If initiated now, the change to a sustainable energy mix leads to less expensive power generation than a business as usual strategy in a time span of about 15 years. Imported fuels with escalating cost will be increasingly substituted by renewable, mostly domestic energy sources. The negative socioeconomic impacts of fossil fuel price escalation can be reversed by 2020 if an adequate political and legal framework is established at time. Feed-in tariffs like the German or Spanish Renewable Energy Acts are very effective instruments for the market introduction of renewables. If tariff additions are subsequently reduced to zero, they can be considered a public investment rather than a subsidy.

- Solar electricity generated by concentrating solar thermal power stations in MENA and transferred to Europe via high voltage direct current transmission can provide firm capacity for base load, intermediate and peaking power, effectively complementing European electricity sources. Starting between 2020 and 2025 with a transfer of 60 TWh/y, solar electricity imports could subsequently be extended to 700 TWh/y in 2050. High solar irradiance in MENA and low transmission losses of 10–15% will yield a competitive import solar electricity cost of around 0.05 €/kWh.
- Carbon emissions can be reduced to 25 % compared to the year 2000; 1% of the European land will be required for the power mix, which is equivalent to the land used at present for transport and mobility.
- European support for MENA for the market introduction of renewables can attenuate the growing pressure on fossil fuel resources that would otherwise result from the economic growth of this region, thus helping indirectly to secure fossil fuel supply also in Europe. The necessary political process could be initiated by a renewable energy partnership and a common free trade area for renewable energies in EUMENA and culminate in a Community for Energy, Water and Climate Security.

The TRANS-CSP study provides a first information base for the design of a political framework that is required to initiate and realize such a scheme (Trieb).

14.8.1.1. Concentrating Solar Power for Seawater Desalination (AQUA-CSP)

The general perception of "solar desalination" today comprises only small-scale technologies for decentralized water supply in remote places, which may be quite important for the development of rural areas, but does not address the increasing water deficits of the quickly growing urban centers of demand. Conventional large-scale desalination is perceived as expensive, energy consuming and limited to rich countries like those of the Arabian Gulf, especially in view of the quickly escalating cost of fossil fuels like oil, natural gas and coal. The environmental impacts of large-scale desalination due to airborne emissions of pollutants from energy consumption and to the discharge of brine and chemical additives to the sea are increasingly considered as critical. For those reasons, most contemporary strategies against a "Global Water Crisis" consider seawater desalination only as a marginal element of supply. The focus of most recommendations lies on more efficient use of water, better account-ability, re-use of waste water, enhanced distribution and advanced irri-gation systems. To this adds the recommendation to reduce agriculture and rather import food from other places. On the other hand, most sources that do recommend seawater desalination as part of a solution to the water crisis usually propose nuclear fission and fusion as an indis-pensable option.

None of the presently discussed strategies include concentrating solar power (CSP) for seawater desalination within their portfolio of possible alternatives. However, quickly growing population and water demand and quickly depleting groundwater resources in the arid regions of the world require solutions that are affordable, secure and compatible with the envi-ronment—in one word: sustainable. Such solutions must also be able to cope with the magnitude of the demand and must be based on available or at least demonstrated technology, as strategies bound to uncertain technical breakthroughs, if not achieved in time, would seriously endanger the whole region.

Renewable energy sources have been accepted worldwide as sustainable sources of energy, and are introduced to the energy sector with an annual growth rate of over 25% per year. From all available energy sources, solar energy is the one that correlates best with the demand for water, because it is obviously the main cause of water scarcity. The resource-potential of concentrating solar power dwarfs global energy demand by several hundred times. The environmental impact of its use has been found to be acceptable, as it is based on abundant, recyclable materials like steel, concrete and glass for the concentrating solar thermal collectors. Its cost is today equivalent to about 50 US$ per barrel of fuel oil (8.8 US$/GJ), and coming down by 10–15 % each time the worldwide installed capacity doubles. In the medium-term by 2020, a cost equivalent to about 20 US$ per barrel (3.5 US$/GJ) will be achieved. In the long-term, it will become one of the cheapest sources of energy, at a level as low as 15 US$ per barrel of oil (2.5 US$/GJ). It can deliver energy "around the clock" for the continuous operation of desalination plants, and is therefore the "natural" resource for seawater desalination.

The AQUA-CSP study analyzes the potential of concentrating solar thermal power technology for large-scale seawater desalination for the urban centers in the Middle East and North Africa (MENA). It provides a comprehensive data base on technology options, water demand, reserves and deficits and derives the short-, medium- and long-term markets for solar powered desalination of 20 countries in the region. The study gives a first information base for a political framework that is required for the initiation and realization of such a scheme. It quantifies the available solar energy resources and the expected cost of solar energy and desalted water, a long-term scenario of integration into the water sector, and quantifies the environmental and socio-economic impacts of a broad dissemination of this concept.

There are several good reasons for the implementation of large-scale concentrating solar powered desalination systems that have been identified within the AQUA-CSP study at hand:

- Due to energy storage and hybrid operation with (bio)fuel, concentrating solar power plants can provide around-the-clock firm capacity that is suitable for large scale desalination either by thermal or membrane processes.
- CSP desalination plants can be realized in very large units up to several 100,000 m³/day.
- Huge solar energy potentials of MENA can easily produce the energy necessary to avoid the threatening freshwater deficit that would otherwise grow from today 50 billion cubic meters per year to about 150 billion cubic meters per year by 2050.
- Within two decades, energy from solar thermal power plants will become the least cost option for electricity (below 4 ct/kWh) and desalted water (below 0.4 €/m³).
- Management and efficient use of water, enhanced distribution and irrigation systems, re-use of wastewater and better accountability are important measures for sustainability, but will only be able to avoid about 50% of the long-term deficit of the MENA region.
- Combining efficient use of water and large-scale solar desalination, over-exploitation of groundwater in the MENA region can, and must, be ended around 2030.
- Advanced solar powered desalination with horizontal drain seabed-intake and nano-filtration will avoid most environmental impacts from desalination occurring today.
- With support from Europe, the MENA countries should immediately start to establish favorable political and legal frame conditions for the market introduction of concentrating solar power technology for electricity and seawater desalination.

The AQUA-CSP study shows a sustainable solution to the threatening water crisis in the MENA region, and describes a way to achieve a balanced, affordable and secure water supply structure for the next generation, which has been overlooked by most contemporary strategic analysis.

Important findings: Only 1–2% of the desert area could supply the whole world with electricity and one square kilometer of desert land using CSP Technology can harvest up to:

- 250 million kWh/year of electricity (250 liter oil/m^2)
- 60 million cubic meters/year of desalted water (6000 liter water/m^2)
- Produced electricity would be transmitted to European and African countries by a super grid of high-voltage direct current cables.
- The expected load losses amount to 3% per 1000 km.

Highways for Renewable Electricity (Supergrid) EUMENA – Trans-Mediterranean High Voltage Direct Current Grid (Figure 14.41):

- Produced electricity would be transmitted to European and African countries by a super grid of high-voltage direct current cables.(Trieb)

There are several installations which are constructed worldwide and demonstrate their practical applications (Figure 14.42):

- The United States has for a long time been the only showcase for modern CSP projects, with about 350 MW capacity.
- It is based on parabolic trough technology operating in the SEGS (Solar Electricity Generation Systems) and
- Installed in the Californian Mojave Desert since the mid-1980s (Figure 14.43).

Figure 14.41 *Highways for the projected Renewable Electricity Supergrid.*

Figure 14.42 *The first Linear Fresnel Solar Steam Generator NOVATEC-BioSol, 1.5 MW, in Spain.*

Figure 14.43 *150 MW of parabolic trough CSP plant at Kramer Junction, California, 2009.*

14.8.2. Solar Oases

In these studies the author has designed the concept of creating renewable energy "Solar Oases" (Figure 14.44) by using alternative energy sources and desalinated sea water for supplying the population in arid regions with food,

Figure 14.44 *Projection of the oases including housing, lakes, food crops and greenhouse facilities.* (See color plate 45.)

energy and water and to establish a fundament for development and to ensure their survival in the future:

- Step one: Sea water desalination using concentrated solar thermal power and advanced desalination technologies (Figure 14.45a).
- Step two: Transport of the water for implementation of farms (Figure 14.45b).

Figure 14.45a *CSP fields.*

Figure 14.45b *Pipeline for desalted water transport.*

- Step three: Establishing communities including:
- Step four: Farms and food production facilities together with green houses, fish lakes, etc.
- Step five: Opportunities for job creation, education, handworks, workshops, and marketing.
- Step six: The vision for sustainable survival of the people in many regions (Figure 14.46).

The concept can be also applied for already existing villages and settlements. Specific modifications are then needed.

The mission is to develop a vision for this and future generations. The concept has been visualized in a video clip which can be seen at www.ifeed. org (El Bassam, 2009) (Figure 14.47).

14.9. THE VATICAN CITY

And suddenly there was light!

The least populated sovereign state in the world, with a population of just 800 people, Vatican City has been declared as the world's most environmentally friendly state with the installation of giant solar power panels. The city reached a record in solar energy production per capita with a figure of 200 watts per inhabitant.

In 2008 SolarWorld AG completed the first solar power plant for the Vatican right next to St. Peter's Cathedral. As from today some 2,394 solar modules generate electricity on the roof of the Papal audience hall (Solar World AG 2010). The solar power plant on the roof of the Paolo VI

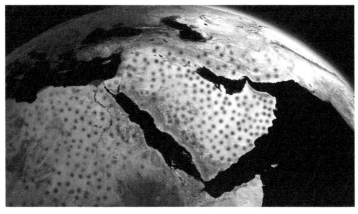

Figure 14.46 *The ultimate vision for distributed solar oases.* (See color plate 46.)

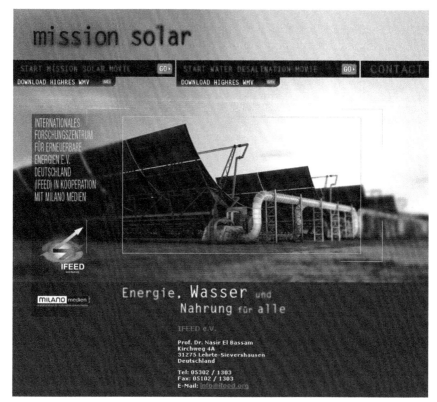

Figure 14.47 *Energy, water and food for survival. www.ifeed.org.*

audience hall has a peak total output of 221.59 kilowatt (kW), enough to generate some 300,000 kilowatt hours of electricity. This is equivalent to the annual needs of more than 100 households. The generation of this volume of clean energy is designed to avoid the emission of 225,000 kilograms of carbon dioxide. The aesthetically sophisticated plant was blended into the historical ensemble of Vatican City with a great deal of technical and architectural effort. The installation of solar panels on the Paul IV conference hall has saved the Vatican 89.84 tons of oil equivalent (Figure 14.49).

The Vatican is not exactly a large state, so its proposed solar plant will generate enough energy to power all of its 40,000 households. The installation will be located on a 740-acre site near Santa Maria di Galeria—the same place where the Vatican Radio's transmission tower is located. The energy that it produces will be far above the needs of the entire Vatican, providing enough power to meet the needs of Vatican radio nine times over.

Figure 14.48 *The Vatican City received first solar power plant in 2008. (Jolly, A 2010. "Vatican City becomes the greenest state in the world," MSN news).*

Figure 14.49 *Powered by sun energy: 2394 Modules, 300 MWh/y, and less 225 t/y/ of Emitted CO2. (Solarworld Group 2008).* (See color plate 47.)

These are not the only moves that the Vatican is taking to reduce its greenhouse emissions. It is contemplating using an electric Pope mobile, the Vatican cafeteria will soon be decked with a solar heating system to provide heating and cooling, and even the Pope's summer residence is being fitted out to get power from the methane generated by the horse stables.

The director of Vatican Radio has declared: "When looking for inspiration, the Pope clearly defers towards the heavens, but when looking for electricity, the sun is his choice."

REFERENCES

Allen, C., 2011. German village achieves energy independence and then some [WWW] BioCycle. Available from http://www.infiniteunknown.net/2011/08/22/german-village-wildpoldsried-pop-2600-produces-321-percent-more-energy-than-it-needs-and-is-generating-5-7-million-in-annual-revenue/.

Bloomberg, L.P., 2012. New Energy Finance [WWW]. Available from http://www.newenergyfinance.com/PressReleases/view/186.

Boston Consulting Group, 2009. Electricity Storage Making Large-Scale Adoption of Wind and Solar Energies a Reality". http://www.bcg.no/expertise_impact/Industries/Energy_Environment/PublicationDetails.aspx?id=tcm:106-41977.

Danish Energy Agency, Heat supply: Goals and means over the years. http://dbdh.dk/images/uploads/pdf-diverse/varmeforsyning%20i%20DK%20p%C3%A5%20engelsk.pdf.

El Bassam, N., 2010. Handbook of bioenergy crops: A complete reference to species development and applications. In: 2nd ed (Ed.). Taylor & Francis Group Ltd, Oxford, Routledge.

El Bassam, N., Maegaard, P., 2004. Integrated renewable energy for rural communities, planning guidelines, technologies and applications. Elsevier, The Netherlands.

Etcheverry, J., 2008. Challenges and opportunities for implementing sustainable energy strategies in coastal communities of Baja California Sur. University of Toronto, Mexico.

Federal Ministry of Food, Agriculture and Consumer Protection, Berlin Office 11055 Berlin [WWW]. Available from http://www.wege-zum-bioenergiedorf.de/metanavigation/datenschutz/.

Iraq Dream Homes, LLC (IDH) 2010 [WWW]. Available from http://www.iraq-homes.com/about_us-en.html.

IZNE 2005 The Bioenergy Village Self-sufficient Heating and Electricity Supply Using Biomass. www.bioenergiedorf.de.

IZNE 2007 Wärme- und Stromversorgung durch heimische Biomasse April 2007, Interdisziplinäres Zentrum für Nachhaltige Entwicklung (IZNE) der Universität Göttingen. - Projektgruppe Bioenergiedörfer. http://www.gar-bw.de/fileadmin/gar/pdf/Energie_und_Klima/Juehnde.pdf.

Lund, H., 2010. Renewable energy systems: The choice and modeling of 100% renewable solutions. Academic Press.

Maegaard, P., 2008. Denmark: Wind Leader in Stand-By. International Sustainable Energy Review. http://xplqa30.ieee.org/xpl/references.jsp?arnumber=5673216.

Maegaard, P., 2009. Danish Renewable Energy Policy, article. WCRE. org homepage September 2009.

Maegaard, P., 2009/2010. Thisted: 100% Renewable Energy municipality. PowerPoint presentation.

Maegaard, P., 2010. Wind Energy Development and Application Prospects of Non-Grid-Connected Wind Power. In: Proceedings of 2009 World Non-Grid-Connected Wind Power and Energy Conference. IEEE Press.

Marshall, M 2012, New Scientist, issue 2850, accessed from http://www.newscientist.com/article/mg21328505.000-indias-panel-price-crash-could-spark-solar-revolution.html.

Northern Territory Government Australia: Roadmap to Renewable and Low Emission Energy in Remote Communities Report, http://www.greeningnt.nt.gov.au/climate/docs/Renewable_Energy_Report_FA.pdf 2011.

Maegaard, Preben, June 2011. Integrated Systems to Reduce Global Warming 2011, Springer Science+Business Media. LLC 22 PDF. http://www.springer.com/about+springer/locations+worldwide?SGWID=4-173904-2052-653447-150.

Ruby, J.U., 2006. Continuous headwind – pioneering the transition from fossil fuels and atomic energy to the renewable energies. Hovedland Danish.

Ruwisch V.B. Sauer 2007. Bioenergy Village Jühnde: Experiences in rural self sufficiency.I, IZNE www.bioenergiedorf.de.

Scheer, H., 2006. Energy autonomy: The economic, social and technological case for renewable energy. Earthscan 2006.

Smeets, E., Faaj, A., Lewandowski, I., 2006. "Progress in Energy and Combustion Science", in A quick scan of global bioenergy potentials to 2050.

El Bassam, N., © 2009, Solar Oases: Transformation of Deserts into Gardens, www.ifeed.org.

SolarWorld Group, 2008. Solar Power for the Vatican: Inauguration of the First Solar Power Plant for the Papal State. Bonn [WWW]. Available from http://solarworld-usa. com/news-and-resources/news/vatican-inauguration-solar-power-for-papal-state.aspx.

Workshop, Strategy, June 2011. Green Settlements in Sub Saharan Africa: Future Land Use and Pathways to Wealth Creation. Loccum Protestant Academy, Germany.

The Poul la Cour Foundation, 2009. Wind power, the Danish way, from Poul la Cour to modern wind turbines. Self-Published, Denmark.

Toke, D., 2007. Evaluation and Recommendations. DESIRE project – Dissemination strategy on Electricity balancing large scale integration of Renewable Energy.

Trieb, Franz, Dr., 2005. MED-CSP, Final Reports. German Aerospace Center (DLR).

Institute of Technical Thermodynamics, Section Systems Analysis and Technology Assessment, Study commissioned by: Federal Ministry for the Environment, Nature Conservation and Nuclear Safety, Germany, Institute of Technical Thermodynamics, Section Systems Analysis and Technology Assessment, www.dlr.de/tt/med-csp.

Trieb, Franz, Dr., 2006. TRANS-CSP. Final Reports. German Aerospace Center (DLR).

Institute of Technical Thermodynamics, Section Systems Analysis and Technology Assessment, Study commissioned by: Federal Ministry for the Environment, Nature Conservation and Nuclear Safety, Germany, Institute of Technical Thermodynamics, Section Systems Analysis and Technology Assessment, www.dlr.de/tt/trans-csp.

Trieb, Franz, Dr., 2007. AQUA-CSP, Final Reports. German Aerospace Center (DLR).

Institute of Technical Thermodynamics, Section Systems Analysis and Technology Assessment, Study commissioned by: Federal Ministry for the Environment, Nature Conservation and Nuclear Safety, Germany, Institute of Technical Thermodynamics, Section Systems Analysis and Technology Assessment, http://www.dlr. de/tt/aqua-csp.

FURTHER READING

Abramsky, K., 2010. Sparking a worldwide energy revolution: Social struggles in the transition to a post-petrol world. AK Press, Oakland.

El Bassam, N., Maegaard, P., 2004. Integrated renewable energy for rural communities, planning guidelines, technologies and applications. Elsevier, The Netherlands.

Etcheverry, J., 2008. Challenges and opportunities for implementing sustainable energy strategies in coastal communities of Baja California Sur. University of Toronto, Mexico.

Lund, H., 2010. Renewable energy systems: The choice and modeling of 100% renewable solutions. Academic Press.

Maegaard, P., 2008. Denmark: Wind Leader in Stand-By. International Sustainable Energy Review.

Maegaard, P., 2009. Danish Renewable Energy Policy, article, WCRE*org homepage. September 2009.

Maegaard, P., 2009/2010. Thisted: 100% Renewable Energy municipality. Powerpoint presentation.

Maegaard, P., 2010. Wind Energy Development and Application Prospects, of Non-Grid-Connected Wind Power. In: Proceedings of 2009 World Non-Grid-Connected Wind Power and Energy Conference. IEEE Press.

Ruby, J.U., 2006. Continuous headwind – pioneering the transition from fossil fuels and atomic energy to the renewable energies. Hovedland. 2006, (in Danish language).

Scheer, H .2006., Energy Autonomyautonomy, : The Economiceconomic, Social social and Technological technological Case case for Renewable renewable Energyenergy, Earthscan, 2006.

Smeets, E., Faaj, A., Lewandowski, I., 2006. Progress in Energy and Combustion Science. In: A quick scan of global bioenergy potentials to 2050.

CHAPTER FIFTEEN

Ownership, Citizens Participation and Economic Trends

15.1. COMMUNITY OWNERSHIP

Interest in community ownership of renewable energy generation is increasing not only in Europe but also in North America, following the launch of Ontario's groundbreaking feed-in tariff program of 2009. Ontario's policy specifically encourages community and aboriginal ownership of renewables. Currently 800 MW of projects, a full 20% of all projects in the Ontario program, are under contract, though not yet built.

The "feed-in" law passed in 1999 has paved the way for electricity generated in Germany, spreading rapidly from Bavaria in the south all the way to the Danish border in the north. In 2010, 51% of the more than 50,000 MW of renewable energy capacity in Germany was owned by farmers, individuals and community groups. This represents a staggering $100 billion in private investment.

German farmers alone have installed 1,600 MW of biogas plants and 3,600 MW of solar photovoltaics (solar PV). For comparison, in 2010 there were only 60 MW of biogas plants and 2,200 MW of solar PV in the entire USA.

German farmers, community leaders and entrepreneurs are not only endorsing electricity generation and renewable heat, but are also setting their sights on an equally ambitious prize, the transmission system itself (Gipe 2012).

15.2. THE DANISH OWNERSHIP MODEL

Non-technical aspects are also of importance for a successful transition to renewable energy supply. The absence of financial investors in the beginning made the wind sector in Denmark unique compared to other countries. At the turn of the century, around 150,000 households were co-owners of a local windmill. The ownership model rather than the technology and tariff schemes were substantial for the success of wind energy

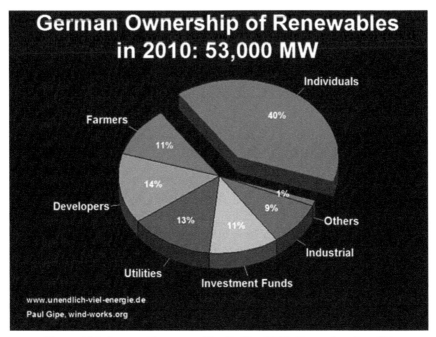

Figure 15.1 *Renewables ownership. (P. Gipe 2012).* (See color plate 48.)

in Denmark. It was the key factor behind the high public acceptance that wind power projects enjoyed during that time. It also enabled a much faster deployment, since large numbers of people were involved in the sector that provided the necessary good will.

In 1992 the role of cooperatives shrank due to a change in planning procedures. In 1998, due to liberalization in the sector, the ownership model changed dramatically. The residential criteria for ownership were abolished, and everyone was allowed to own a number of windmills as they could get building permission for anywhere in the country. The take-over bids began, resulting in a dramatic decrease in public involvement. As a result, in 2009 about 50,000 households were co-owners of windmills compared to 150,000 a decade earlier.

Consequently, the attitude toward wind power suffered a reversal. The erection of a windmill often became a local drama and resulted in bitter conflicts that led to long delays or cancellations. In 2008 a wind turbine neighbor compensation scheme was introduced that also caused local conflicts. The solution seemed to be to "normalize" the wind energy sector so that wind power like district heating, CHP and power distribution

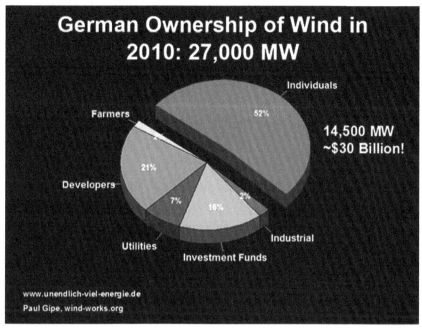

Figure 15.2 *Wind energy ownership.* (*P. Gipe 2012*). (See color plate 49.)

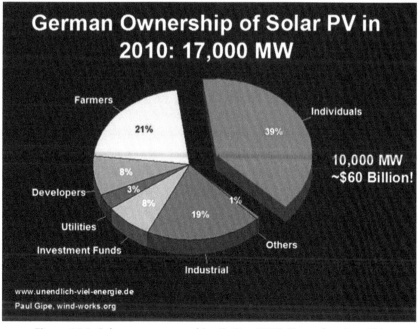

Figure 15.3 *Solar energy ownership.* (*P. Gipe 2012*). (See color plate 50.)

changed from investment to public supply, with municipalities and consumer-owned local companies as the future wind power owners. This made wind power significantly cheaper.

Denmark already has hundreds of consumer-owned local energy supply companies for combined heat and power, district heating and power distribution. New organizational structures and nonprofit ownership models to the direct benefit of the involved municipalities may prove to be the most realistic long-term solution for community wind power as well, to again make it locally acceptable.

15.2.1. Integration of the Energy Supply by Public Ownership

Renewable energy is still young; technology and tariff problems have found reasonable solutions in generally acceptable organizational structures for decentralized ownership for the common good as well as part of the transition process.

The ownership and operation of large wind turbines for community supply should be a service provided by the local public authorities like the CHP, supply of water, central district heating, public transport and other parts of the public infrastructure. It is the lesson from the past 100 years of practice that the state should promote public regulation in favor of local and collective ownership of basic public services, such as energy. This approach is in line with the protection and promotion of the common good in most democratic societies. Considering the size of order and complexity of the transition to a 100% renewable energy-based energy system, and its urgency, it is only realistic for public administrations to undertake this task.

The common good approach will make wind power a part of the public planning with expropriation of the necessary areas for wind turbines, as is done with power pylons, waterworks and similar areas of public interest. It is already standard practice to provide monetary compensation when areas are being designated for the common good. This would be a decisive contribution for attaining local acceptance and make wind energy more competitive. In Denmark it could lead to a 30% cost reduction or more than can be expected from more efficient technology.

In several countries new legislation instructs utilities to buy the electricity from renewable energy installations at a price determined by the government and guaranteed for 20 years. Such laws are used in the most successful wind energy countries. This would be a serious incentive for the municipalities as well, to actively be a part of the development of CHP combined with renewable energy. This can be done through municipal companies or existing and eventually new local renewable supply companies that can supply a full renewable energy package including wind power.

By making the establishment and operation of large wind turbines the responsibility of public supply companies, there will be significant savings due to cheaper areas for wind turbines, saved repowering fees and cheaper long term financing. This will make wind energy more attractive for the individual municipality as well as at a national level and improve supply security, secure steady energy prices and be the fulfillment of international agreements concerning CO_2 reduction as well.

15.3. ECONOMIC TRENDS

Renewable energy commercialization involves the deployment of three generations of renewable energy technologies dating back more than 100 years. First-generation technologies, which are already mature and economically competitive, include biomass, hydroelectricity, geothermal power and heat. Second-generation technologies are market-ready and are being deployed at the present time; they include solar heating, photovoltaics, wind power, solar thermal power stations, and modern forms of bioenergy. Third-generation technologies require continued R&D efforts in order to make large contributions on a global scale and include advanced biomass gasification, bio-refinery technologies, hot-dry-rock geothermal power, and ocean energy (International Energy Agency 2007).

There are some non-technical barriers to the widespread use of renewables (National Renewable Energy Laboratory (2006)), and it is often public policy and political leadership that help to address these barriers and drive the wider acceptance of renewable energy technologies (Aitken 2010). As of 2010, 98 countries have targets for their own renewable energy futures, and have enacted wide-ranging public policies to promote renewables (REN21 2012). Climate change concerns are driving increasing growth in the renewable energy industries (United Nations Environment Programme and New Energy Finance Ltd. 2007). Leading renewable energy companies include First Solar, Gamesa, GE Energy, Q-Cells, Sharp Solar, Siemens, SunOpta, Suntech, and Vestas (Johnson 2010).

Total investment in renewable energy reached $211 billion in 2010, up from $160 billion in 2009. The top countries for investment in 2010 were China, Germany, the United States, Italy, and Brazil. Continued growth for the renewable energy sector is expected and promotional policies helped the industry weather the 2009 economic crisis better than many other sectors.

Globally, there are an estimated 3 million direct jobs in renewable energy industries, with about half of them in the biofuels industry (REN21 2010). According to a 2011 projection by the International Energy Agency, solar power generators may produce most of the world's electricity within 50 years, dramatically reducing harmful greenhouse gas emissions that harm the environment (Sills 2011).

Photovoltaic and solar-thermal plants may meet most of the world's demand for electricity by 2060 and half of all energy needs, with wind, hydropower and biomass plants supplying much of the remaining.

A 2011 IEA report said: "A portfolio of renewable energy technologies is becoming cost-competitive in an increasingly broad range of circumstances, in some cases providing investment opportunities without the need for specific economic support," and added that "cost reductions in critical technologies, such as wind and solar, are set to continue (Gloystein 2011). As of 2011, there have been substantial reductions in the cost of solar and wind technologies."

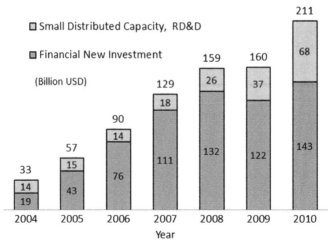

Figure 15.4 *Global new investments in renewable energy. (Data source: Bloomberg New energy Finance, UNEP SEFI, Frankfurt School, Global Trends in Renewable Energy Investment 2011).*

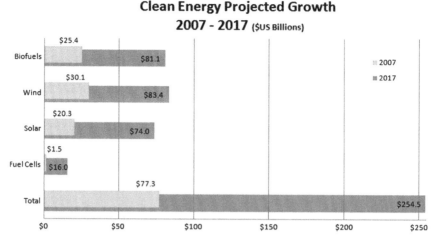

Figure 15.5 *Clean energy projected growth 2007–2017*. *(GGByte 2008, from Clean Energy Trends Report).*

The price of PV modules per MW has fallen by 60 percent since the summer of 2008, according to Bloomberg New Energy Finance estimates (2011), putting solar power for the first time on a competitive footing with the retail price of electricity in a number of sunny countries. Wind turbine prices have also fallen—by 18 percent per MW in the last two years—reflecting, as with solar, fierce competition in the supply chain. Further improvements in the leveled cost of energy for solar, wind and other technologies lie ahead, posing a growing threat to the dominance of fossil fuel generation sources in the next few years *(Renewable Energy World 2011)*.

The International Solar Energy Society argues that renewable energy technologies and economics will continue to improve with time, and that they are "sufficiently advanced at present to allow for major penetrations of renewable energy into the mainstream energy and societal infrastructures." (Aitken 2010). Indicative, leveled, economic costs for renewable power (exclusive of subsidies or policy incentives) are shown in Table 15.1.

As time progresses, renewable energy will generally get cheaper (Analysis 2009), while fossil fuels will generally get more expensive (Gore 2009).

First, once the renewable infrastructure is built, the fuel is free forever. Unlike carbon-based fuels, the wind and the sun and the Earth itself provide fuel that is free, in amounts that are effectively limitless.

Second, while fossil fuel technologies are more mature, renewable energy technologies are being rapidly improved. So innovation and

Table 15.1 (REN21 2010) Renewable power generation costs 2010

Power generator	Typical characteristics	Typical electricity costs (U.S. cents/kWh)
Large hydro	Plant size: 10—18,000 MW	3—5
Small hydro	Plant size: 1—10 MW	5—12
Onshore wind	Turbine size: 1.5—3.5 MW	5—9
Offshore wind	Turbine size: 1.5—5 MW	10—14
Biomass power	Plant size: 1—20 MW	5—12
Geothermal power	Plant size: 1—100 MW	4—7
Rooftop solar PV	Peak capacity: 2—5 kilowatts-peak	20—50
Utility-scale solar PV	Peak capacity: 200 kW to 100 MW	15—30
Concentrating solar thermal power (CSP)	50—500 MW trough	14—18

ingenuity give us the ability to constantly increase the efficiency of renewable energy and continually reduce its cost.

Third, once the world makes a clear commitment to shifting toward renewable energy, the volume of production itself will sharply reduce the cost of each windmill and each solar panel, while adding yet more incentives for additional research and development to further speed up the innovation process (Gore 2009).

REFERENCES

Aitken, Donald W., January 2010. Transitioning to a Renewable Energy Future. International Solar Energy Society, p. 3.

Bloomberg New Energy Finance, 2011. Global New Investments in Renewable Energy UNEP SEFI, Frankfurt School, Global Trends in Renewable Energy Investment. Clean energy projected growth 2007-2017.

Clean Edge (2008) Energy Trends Report (GGByte at en.wikipedia 2012) http://en.wikipedia.org/wiki/File:Re_investment_2007-2017.jpg.

Gipe, P., 2012. From data by Unendlich viel energie. Windworks. January 5, 2012 [WWW]. Available from http://wind-works.org/coopwind/CitizenPowerConferencetobeheld inHistoricChamber.html.

Gipe, P., 2012. Germany renewable energy generation is owned by its own citizens. Windworks [WWW], January 5, 2012. http://wind-works.org/coopwind/Citizen PowerConferencetobeheldinHistoricChamber.html.

Gloystein, Henning, November 23, 2011. "Renewable energy becoming cost competitive, IEA says". Reuters.

Gore, Al, 2009. Our Choice, p. 58. Bloomsbury Publishing. PLC, ISBN: 0747590990.

International Energy Agency (2007).

Johnson, Keith. Wind Shear: GE Wins, Vestas Loses in Wind-Power Market Race. Wall Street Journal March 25 2009, accessed on January 7 2010.

National Renewable Energy Laboratory, 2006. Nontechnical Barriers to Solar Energy Use: Review of Recent Literature. Technical Report NREL/TP-520-40116, September, 30 pages.

REN21, 2010. Renewables 2010 Global Status Report, pp. 9, 26 & 34.

REN21, 2011. "Renewables 2011: Global Status Report", pp. 11–13.

REN21, 2012. Renewables Global Status Report, 2012, p. 17.

Renewables in global energy supply: An IEA facts sheet (PDF) OECD, 34 pages.

"Renewables Investment Breaks Records", August 2011. Renewable Energy World 29.

Sills, Ben, August 29, 2011. Solar May Produce Most of World's Power by 2060, IEA Says. Bloomberg.

Solar Power 50% Cheaper By Year End - Analysis Reuters, November 24, 2009.

United Nations Environment Programme and New Energy Finance Ltd, 2007. Global Trends in Sustainable Energy Investment. 2007. Analysis of Trends and Issues in the Financing of Renewable Energy and Energy Efficiency in OECD and Developing Countries, 3 (PDF).

GLOSSARY

REGIONAL DEFINITIONS

In this Statement the world is divided into two main groups:

- the market-oriented economies (both developed and developing), and
- the economies in transition (which include both industrialized and developing economies).

The developed market economies are, with the exception of Turkey, those of the OECD at the time when ETW was published in 1993. They include North America (US and Canada), Iceland, Western Europe (including the former Eastern part of Germany but excluding Turkey) and the Pacific countries (Japan, Australia and New Zealand). It should be noted that the OECD now includes three economies in transition (the Czech Republic, Hungary and Poland) and two developing countries (the Republic of Korea and Mexico).

The developing market economies include Latin America (which includes Mexico and the Caribbean), Africa, and Asia (including Turkey and the Middle East but excluding the economies in transition of Asia).

The developed economies in transition are the countries of Central and Eastern Europe (CEE), the Commonwealth of Independent States (CIS), and other republics of the Former Soviet Union. We recognize that some of these dividing lines, for instance in Central Asia, between industrialized and developing economies are extremely fine.

The developing economies in transition include China, Cambodia, DPR of Korea, DPR of Laos, Vietnam and Mongolia.

WEC now uses the following regional divisions:

- Africa, which includes North Africa from Algeria to Egypt.
- Asia, which includes Asia Pacific, Middle East and West Asia (including Turkey), and South Asia.
- Europe, which includes Central and Eastern Europe, the CIS, and Western Europe.
- Latin America and the Caribbean (including Mexico).
- North America (US and Canada).

Accuracy The degree to which the mean of a sample approaches the true mean of the population; lack of bias.

Activity A practice or ensemble of practices that take place on a delineated area over a given period of time.

Adaptation Adjusting natural or human systems to cope with actual or expected climate change and its impacts.

Additives Chemicals added to fuel in very small quantities to improve and maintain fuel quality. Detergents and corrosion inhibitors are examples of gasoline additives.

Additionality A project is *additional* if it would not have happened, but for the incentive provided by the credit trading program (e.g., CDM or JI). The Kyoto Protocol specifies that only projects that provide emission reductions that are *additional* to any that would occur in the absence of the project activity shall be awarded certified emission reductions (CERs) in the case of CDM projects or emission reduction units (ERUs) in the case of JI projects. This is often referred to as "environmental additionality."

Advanced Technology Vehicle (ATV) A vehicle that combines new engine/power/drive train systems to significantly improve fuel economy. This includes hybrid power systems and fuel cells, as well as some specialized electric vehicles.

Agriculture, Small Scale Ways to improve crop yields in small areas; permaculture and proven, new techniques.

Agriculture, Larger Scale New farming methods for high altitude and specific growing conditions.

Air Toxics Toxic air pollutants, including benzene, formaldehyde, acetaldehyde, 1-3 butadiene and polycyclic organic matter (POM). Benzene is a constituent of motor vehicle exhaust, evaporative and refueling emissions. The other compounds are exhaust pollutants.

Alcohols Organic compounds that are distinguished from hydrocarbons by the inclusion of a hydroxyl group. The two simplest alcohols are methanol and ethanol.

Aldehydes A class of organic compounds derived by removing the hydrogen atoms from an alcohol. Aldehydes can be produced from the oxidation of an alcohol.

Alternative fuel Methanol, denatured ethanol, and other alcohols; mixtures containing 85% or more by volume of methanol, denatured ethanol, and other alcohols with gasoline or other fuels; natural gas; liquefied petroleum gas; hydrogen; coal-derived liquid fuels; non-alcohol fuels (such as biodiesel) derived from biological material; and electricity.

Alternative energy Energy derived from non-fossil fuel sources.

American Society for Testing and Materials (ASTM) A nonprofit organization that provides a management system to develop published technical information. ASTM standards, test methods, specifications and procedures are recognized as definitive guidelines for motor fuel quality as well as a broad range of other products and procedures.

Ancillary effects Side effects of policies to reduce net greenhouse gas emissions, such as reductions in air pollutants associated with fossil fuels or socioeconomic impacts on employment or agricultural efficiency.

Anhydrous Describes a compound that does not contain any water. Ethanol produced for fuel use is often referred to as anhydrous ethanol, as it has had almost all water removed.

Animal husbandry Taking care of animals, improving strains, basic veterinary skills.

Anthropogenic Manmade.

Anthropogenic emissions Greenhouse gas emissions associated with human activities such as burning fossil fuels or cutting down trees.

Annex I Parties Originally 35 countries, comprising industrialized economies (including Turkey, which has since sought to withdraw voluntarily from the Climate Convention) and 11 transitional economies plus the then European Economic Community as a regional organization, that are Parties to the UN Framework Convention on Climate Change and so listed in Annex I to the Convention. Since 1992 the withdrawal of Turkey and the creation of separate Czech and Slovakian Republics, Croatia and Slovenia, have modified the number of countries.

Annex B Parties Thirty-nine countries listed in Annex B to the Kyoto Protocol, which indicated agreement at the Third Conference of the Parties to the UN Climate Convention to contemplate legally binding quantified emission limitation and reduction commitments. The Kyoto Protocol has not yet been ratified.

Aquaculture Renewable energy powered fish farming for nutrition, health, help in ending hunger and micro-enterprise.

Aromatics Hydrocarbons based on the ringed six-carbon benzene series or related organic groups. Benzene, Toluene and Xylene are the principal aromatics, commonly referred to as the BTX group. They represent one of the heaviest fractions in gasoline.

Array Often several modules are wired together in an array to increase the total available power output and to match the operating voltage of the "load" equipment.

Balance of payments The dollar amount difference between a country's exports and imports. In the United States, large oil imports are one of the main causes of the negative balance of payments with the rest of the world.

Barrier Any obstacle to the diffusion of cost-effective mitigation technologies or practices, whether institutional, social, economic, political, cultural or technological.

Baseline The greenhouse gas emissions level that would occur in the absence of climate change interventions; used as a basis for analyzing the effectiveness of mitigation policies.

Baseline scenario A baseline scenario is a presumed counterfactual alternative to the proposed project. In other words, it is an interpretation of "what would have happened otherwise." Several plausible baseline scenarios can be evaluated for a given project. The project itself can and should typically be considered as one of these baseline scenarios, since the possibility it would have been implemented in the absence of carbon credits must be examined to determine whether it is additional.

Baseline emission rate A baseline emission rate is the parameter, expressed in t CO_2/MWh for electricity projects, which is used to calculate the number of emission credits (i.e., CERs or ERUs) a project can generate. The baseline emissions rate can be based on standardized (multi-project) methodologies, or correspond directly to a project specific baseline scenario.

Baseline scenario analysis This commonly used baseline methodology involves a process of elaborating, then culling through, a series of plausible baseline scenarios, often involving investment or barrier analysis. This analysis is useful to demonstrate the *additionality* of a project.

Base load That part of total energy demand that does not vary over a given period of time.

Benzene A six-carbon aromatic; common gasoline component identified as being toxic. Benzene is a known carcinogen.

Bias Systematic over- or under-estimation of a quantity.

Biochemical Conversion The use of enzymes and catalysts to change biological substances chemically to produce energy products. For example, the digestion of organic wastes or sewage by microorganisms to produce methane is a biochemical process.

Biodiesel A biodegradable transportation fuel for use in diesel engines that is produced through transesterification of organically derived oils or fats. Biodiesel is used as a substitute of diesel fuel.

Biofuels Fuel produced from dry organic matter or combustible oils from plants, such as alcohol from fermented sugar, black liquor from the paper manufacturing process, wood, and soybean oil.

Biogas Methane rich gas containing 30–45% CO_2 and various trace elements, especially H2S. Biogas is produced from organic wastes and residues from agriculture and households by anaerobic fermentation in digesters of sizes from 2 to 3000 cubic meters and temperatures between 20 and 50 C.

Bio-generator power system Village-scale power in remote locations without a fuel source, much less expensive than solar or wind, uses available agricultural waste.

Biological options There are three conserving an existing carbon pool, and thereby preventing emissions into the atmosphere; sequestrating more CO_2 from the atmosphere by increasing the size of existing carbon pools; and substituting biological products for fossil fuels or for energy-intensive products, thereby reducing CO_2 emissions.

Biomass The total mass of living organisms in a given area or volume; biomass can be used as a sustainable source of fuel with low or zero net emissions.

Biosphere That component of the Earth system that contains life in its various forms, which includes its living organisms and derived organic matter (e.g., litter, detritus, soil).

British Thermal Unit (Btu) A standard unit for measuring heat energy. One Btu represents the amount of heat required to raise one pound of water one degree Fahrenheit (at sea level).

Briquetting Making a more efficient, smokeless fuel source with local materials and without wood, keeping dung for gardens and pastures instead of going up in smoke.

BTX Industry term referring to the group of aromatic hydrocarbons benzene, toluene and xylene (see Aromatics).

Build margin The build margin refers to new sources of electric capacity expected to be built or otherwise added to the system, and affected by a new project-based activity.

Butane A gas, easily liquefied, recovered from natural gas. Used as a low-volatility component of motor gasoline, processed further for a high-octane gasoline component, used in LPG for domestic and industrial applications and used as a raw material for petrochemical synthesis.

Butyl alcohol Alcohol derived from butane that is used in organic synthesis and as a solvent.

Carbon dioxide (CO_2) The gas formed in the ordinary combustion of carbon, given out in the breathing of animals, burning of fossil fuels, etc. Human sources are very small in relation to the natural cycle.

Carbon monoxide (CO) A colorless, odorless gas produced by the incomplete combustion of fuels with a limited oxygen supply, as in automobile engines.

Carbon flux Transfer of carbon from one carbon pool to another in units of measurement of mass per unit area and time (e.g., $t\ C\ ha^{-1}\ y^{-1}$).

Carbon pool A reservoir. A system that has the capacity to accumulate or release carbon. Examples of carbon pools are forest biomass, wood products, soils, and atmosphere. The units are mass (e.g., t C).

Carbon sequestration The absorption and storage of CO_2 from the atmosphere by the roots and leaves of plants; the carbon builds up as organic matter in the soil.

Carbon stock An absolute quantity of carbon held within a pool at a specified time.

Carcinogens Chemicals and other substances known to cause cancer.

Catalyst A substance whose presence changes the rate of chemical reaction without itself undergoing permanent change in its composition. Catalysts may be accelerators or retarders. Most inorganic catalysts are powdered metals and metal oxides, chiefly used in the petroleum, vehicle and heavy chemical industries.

Cetane Ignition performance rating of diesel fuel. Diesel equivalent to gasoline octane.

Closed-loop carburetion System in which the fuel/air ratio in the engine is carefully controlled to optimize emissions performance. A closed-loop system uses a fuel metering correction signal to optimize fuel metering.

Co-solvents Heavier molecular weight alcohols used with methanol to improve water tolerance and reduce other negative characteristics of gasoline/alcohol blends.

Combined margin baselines A combined margin baseline reflects both operating and build margin effects.

Communications Bridging the "digital divide"; satellite, long-range phone, and Internet communication systems.

Commercial energy Energy supplied on commercial terms. Distinguished from non-commercial energy comprising fuel wood, agricultural wastes and animal dung collected usually by the user.

Community organization Mobilizing community involvement, organizing a task-force, and developing a five-year village development plan.

Compressed natural gas (CNG) Natural gas that has been compressed under high pressures, typically 2,000 ue metric to 3,600 psi, held in a container. The gas expands when used as a fuel.

Compression ignition of diesel engine The form of ignition that initiates combustion in a diesel engine. The rapid compression of air within the cylinders generates the heat required to ignite the fuel as it is injected.

Converted or conversion vehicle A vehicle originally designed to operate on gasoline or diesel that has been modified or altered to run on an alternative fuel.

Cooking oil production Hand and solar-powered presses to make everything from cooking oil to bio-fuels that can power vehicles.

Cooling (refrigeration) systems How to make and use "pot-within-a-pot" earthenware cooling systems with locally available materials (clay pots, sand and cloth). In dry climates, using these super-inexpensive systems, vegetables stay fresh for weeks instead of just a few days.

Corrosion inhibitors Additives used to inhibit corrosion in the fuel system (e.g., rust).

Cryogenic storage Extreme low-temperature storage.

Dedicated natural gas vehicle A vehicle that operates only on natural gas. Such a vehicle is incapable of running on any other fuel.

Dedicated vehicle A vehicle that operates solely on one fuel. Generally, dedicated vehicles have superior emissions and performance results because their design has been optimized for operation on a single fuel.

Denatured alcohol Ethanol that contains a small amount of a toxic substance, such as methanol or gasoline, which cannot be removed easily by chemical or physical means. Alcohols intended for industrial use must be denatured to avoid federal alcoholic beverage tax.

Detergent Additives used to inhibit deposit formation in the fuel and intake systems in automobiles.

Direct emissions Direct emissions are a direct consequence of project activity, either on-site (e.g., via fuel combustion at the project site) or off-site (e.g., from grid electricity or district heat, and other upstream and downstream life cycle impacts).

Distillation curve The percentages of gasoline that evaporate at various temperatures. The distillation curve is an important indicator for fuel standards such as volatility (vaporization).

Distributed generation This term is used for a generating plant serving a customer on-site, or providing support to a distribution network, and connected to the grid at distribution level voltages. The technologies generally include engines, small (including micro) turbines, fuel cells, and photovoltaics.

Dual-fuel vehicle Vehicle designed to operate on a combination of an alternative fuel and a conventional fuel. This includes a) vehicles using a mixture of gasoline or diesel and an alternative fuel in one fuel tank, commonly called flexible-fueled vehicles; and b) vehicles capable of operating either on an alternative fuel, a conventional fuel or both, simultaneously using two fuel systems commonly called bi-fuel vehicles.

Dynamometer An instrument for measuring mechanical force, or an apparatus for measuring mechanical power (as of an engine).

E10 (Gasohol) Ethanol mixture containing 93% ethanol, 5% methanol and 2% kerosene, by volume.

E85 Ethanol/gasoline mixture containing 85% denatured ethanol and 15% gasoline, by volume.

E93 Ethanol mixture containing 93% ethanol, 5% methanol and 2% kerosene, by volume.

E95 Ethanol/gasoline mixture containing 95% denatured ethanol and 5% gasoline, by volume.

Eco-tourism Ways to develop an income by serving tourists; clean water, showers, clean beds, Internet service.

Education Training the trainers; educational methodologies for assuring the best participation and success of project implementation.

Electric vehicle A vehicle powered by electricity, generally provided by batteries. EVs qualify in the zero emission vehicle (ZEV) categories for emissions.

Electricity Electric current used as a power source. Electricity can be generated from a variety of feedstock including oil, coal, nuclear, hydro, natural gas, wind, and solar. In electric vehicles, onboard rechargeable batteries power an electric motor.

El Niño A warm ocean surge in the East Pacific off the coast of Peru which has occurred every few years for thousands of years, and which regularly causes major climatic disruption over a wide area. Often referred to as an ENSO (El Niño Southern Oscillation) event. The reverse phase of El Niño is referred to as La Niña.

Emissions tax A levy imposed by a government on each unit of CO_2 equivalent emissions from a source subject to the tax; can be imposed as a carbon tax to reduce carbon dioxide emissions from fossil fuels.

Emissions trading A market-based approach to achieving environmental objectives that allows countries or companies that reduce greenhouse gas emissions below their target to sell their excess emissions credits or allowances to those that find it more difficult or expensive to meet their own targets.

Energy intensity The proportion of energy used to Gross Domestic Product at constant prices.

Ester An organic compound formed by reacting an acid with an alcohol, always resulting in the elimination of water.

Ethane (C₂H₆) A colorless hydrocarbon gas of slight odor having a gross heating value of 1,773 BT. It is a normal constituent of natural gas.

Ethanol (also known as Ethyl alcohol, Grain alcohol, CH 3 CH 2 OH) Can be produced chemically from ethylene or biologically from the fermentation of various sugars from carbohydrates found in agricultural crops and cellulosic residues from crops or wood. Used in the United States as a gasoline octane enhancer and oxygenate, it increases octane 2.5 to 3.0 numbers at 10% concentration. Ethanol also can be used in higher concentration in alternative-fuel vehicles optimized for its use

Ether A class of organic compounds containing an oxygen atom linked to two organic groups.

Etherification Oxygenation of an olefin by methanol or ethanol. For example, MTBE is formed from the chemical reaction of isobutylene and methanol.

Ethyl alcohol See Ethanol.

Ethyl ester A fatty ester formed when organically derived oils are combined with ethanol in the presence of a catalyst. After water washing, vacuum drying and filtration, the resulting ethyl ester has characteristics similar to petroleum-based diesel motor fuels.

Ethyl Tertiary Butyl Ether (ETBE) A fuel oxygenate used as a gasoline additive to increase octane and reduce engine knock.

Evaporative emissions Hydrocarbon vapors that escape from a fuel storage tank or a vehicle fuel tank or vehicle fuel system.

Flexible-fuel vehicle Vehicles with a common fuel tank designed to run on varying blends of unleaded gasoline with either ethanol or methanol.

Fluidized beds Beds of burning fuel and non-combustible particles kept in suspension by upward flow of combustion air through the bed. Limestone or coal ashes are widely used non-combustible materials.

Flux See Carbon Flux.

Food storage Solar crop drying methods, butter/cheese storage, meat preservation; coffee, nuts, grain.

Forest stand A community of trees, including aboveground and belowground biomass and soils, sufficiently uniform in species composition, age, arrangement, and condition to be managed as a unit.

Fossil fuels Carbon-based fuels from fossil carbon deposits, including coal, oil, and natural gas.

Fuel cell An electrochemical engine with no moving parts that converts the chemical energy of a fuel, such as hydrogen, and an oxidant, such as oxygen, directly to electricity. The principal components of a fuel cell are catalytically activated electrodes for the fuel (anode) and the oxidant (cathode) and an electrolyte to conduct ions between the two electrodes.

Gasifiers Tank for anaerobic fermentation of biomass residues from sugar cane, pulp and paper, etc., to produce biogas.

Gasoline Gallon Equivalent (gge) A unit for measuring alternative fuels so that they can be compared with gasoline on an energy equivalent basis. This is required because the different fuels have different energy densities.

Geothermal Natural heat extracted from the earth's crust using its vertical thermal gradient, most readily available where there is a discontinuity in the earth's crust (e.g., where there is separation or erosion of tectonic plates).

Global warming The theoretical escalation of global temperatures caused by the increase of greenhouse gas emissions in the lower atmosphere.

Greenhouse effect A warming of the earth and its atmosphere as a result of the thermal trapping of incoming solar radiation by CO_2, water vapor, methane, nitrous oxide, chlorofluorocarbons, and other gases, both natural and man-made.

Greenhouse gases Gases which, when concentrated in the atmosphere, prevent solar radiation trapped by the Earth and re-emitted from its surface from escaping. The result is *ceteris paribus*, a rise in the Earth's near surface temperature. The phenomenon was first described by Fourier in 1827, and first termed the greenhouse effect by Arrhenius in 1896. Carbon dioxide is the largest in volume of the greenhouse gases. The others are halocarbons, methane, nitrous oxide, hydro fluorocarbons, perfluorocarbons, and sulphur hexafluoride.

Gross Vehicle Weight (gvw) Maximum weight of a vehicle, including payload.

Halocarbons A family of chlorofluorocarbons (CFCs) mostly of industrial origin—CH_3Cl is the main exception. Includes aerosol propellants (CFCs 11, 12, 114); refrigerants (CFCs 12, 114 and HCFC-22); foam-blowing agents (CFCs 11 and 12); solvents (CFC-113, CH_3CCl_3 and CCl_4); and fire retardants (halons 1211 and 1301). HCFC = hydrofluorocarbon.

Heterotrophic respiration The release of carbon dioxide from decomposition of organic matter.

Hybrid Electric Vehicle A vehicle powered by two or more energy sources, one of which is electricity. HEVs may combine the engine and fuel of a conventional vehicle with the batteries and electric motor of an electric vehicle in a single drive train.

Indirect emissions Indirect emissions occur when market and individual response to a project activity leads to increased or decreased emissions. Indirect emissions can be either on-site (e.g., rebound effects such as increased heating that may result from of an insulation program) or off-site (e.g., project effects that are often referred to as leakage, either negative or positive, such as economy-wide response to price changes or increased penetration of low carbon technologies outside the project site induced by the project activity).

Internet café Ways to create micro-enterprise services from a satellite Internet connection.

Investment additionality Investment additionality seeks to compare the financial return of a project and its alternative (baseline scenario) to determine additionality of a CDM or JI project and/or the most likely baseline.

Land cover The observed physical and biological cover of the Earth's land as vegetation or manmade features.

Land use The total of arrangements, activities, and inputs undertaken in a certain land cover type (a set of human actions). The social and economic purposes for which land is managed (e.g., grazing, timber extraction, conservation).

Lead See Tetraethyl Lead.

Leakage Occurs when emissions reductions in developed countries are partly off-set by increases above baseline levels in developing countries, due to relocation of energy-intensive production, increased consumption of fossil fuels when decreased developed country demand lowers international oil prices, or changes in incomes and thus in energy demand because of better terms of trade, or when sink activities such as tree planting on one parcel of land encourage emitting activities elsewhere.

LED / Hand-generator manufacturing The least expensive lighting systems for remote villages, a proven, small-scale manufacturing opportunity.

LNG to CNG station A station, supplied with LNG, that pumps and vaporizes the liquid supply to vehicles as CNG fuel, generally at the correct pressure and temperature (i.e., the temperature effect of compression is factored into the design).

LNG vehicle A vehicle that uses LNG as its fuel.

Light-duty vehicle Passenger cars and trucks with a gross vehicle weight rating of 3,500 kilograms or less.

Liquefied Natural Gas (LNG) Compressed natural gas that is cryogenically stored in its liquid state.

Liquefied Petroleum Gas (LPG) A mixture of hydrocarbons found in natural gas and produced from crude oil, used principally as a feedstock for the chemical industry, home heating fuel, and motor vehicle fuel. Also known by the principal constituent propane.

Liter (L) A metric measurement used to calculate the volume displacement of an engine. One liter is equal to 1,000 cubic centimeters or 61 cubic inches.

Lubricity Capacity to reduce friction.

M100 100% (neat) methanol.

M85 85% methanol and 15% unleaded gasoline by volume, used as a motor fuel in FFVs.

Methane (CH_4) A gas emitted from coal seams, natural wetlands, rice paddies, enteric fermentation (gases emitted by ruminant animals), biomass burning, anaerobic decay or organic wastes in landfill sites, gas drilling and venting, and the activities of termites.

Methanol (also known as Methyl Alcohol, Wood Alcohol, CH_3, OH) A liquid fuel formed by catalytically combining CO with hydrogen in a 1 to 2 ratio under high temperature and pressure. Commercially, it is typically manufactured by steam reforming natural gas. Also formed in the destructive distillation of wood.

Methyl Alcohol See Methanol.

Methyl Ester A fatty ester formed when organically derived oils are combined with methanol in the presence of a catalyst. Methyl Ester has characteristics similar to petroleum-based diesel motor fuels.

Methyl Tertiary Butyl Ether (MTBE) A fuel oxygenate used as an additive to gasoline to increase octane and reduce engine knock.

Mobile source emissions Emissions resulting from the operations of any type of motor vehicle.

Motor octane The octane as tested in a single-cylinder octane test engine at more severe operating conditions. Motor Octane Number (MON) affects high-speed and part-throttle knock and performance under load, passing, climbing and other operating conditions. Motor octane is represented by the designation M in the (R+M)/2 equation and is the lower of the two numbers.

Micro-finance How to set up and organize a micro-lending organization.

Micro-hydro manufacturing How to make small hydro units from car alternators.

Mitigation Action to reduce sources or enhance sinks of greenhouse gases.

Module A PV module is composed of PV cells that are encapsulated (typically between two glass covers or between a glass front and a suitably weatherproof backing) to protect them against the elements. Such modules, usually framed, are the basic building blocks of PV power systems.

Mushroom growing Ways to quickly improve nutrition. Medicinal benefits and mushrooms as a significant tool for soil restoration, replenishment and remediation.

Natural gas A mixture of gaseous hydrocarbons, primarily methane, occurring naturally in the earth and used principally as a fuel.

Natural gas distribution system This term generally applies to mains, services, and equipment that carry or control the supply of natural gas from a point of local supply, up to and including the sales meter.

Natural gas transmission system Pipelines installed for the purpose of transmitting natural gas from a source or sources of supply to one or more distribution centers.

Near neat fuel Fuel that is virtually free from admixture or dilution.

Neat alcohol fuel Straight or 100% alcohol (not blended with gasoline), usually in the form of either ethanol or methanol.

Neat fuel Fuel that is free from admixture or dilution with other fuels.

No regrets policy Policies that would generate net social benefits whether or not there is climate change; for example, the value of reduced energy costs or local pollution may exceed the costs of cutting the associated emissions.

Non-road vehicle (off-road vehicle) A vehicle that does not travel streets, roads, or highways. Such vehicles include construction vehicles, locomotives, forklifts, tractors, golf carts, and so forth.

OEM Original Equipment Manufacturer.

Octane enhancer Any substance such as MTBE, ETBE, toluene and xylene that is added to gasoline to increase octane and reduce engine knock.

Octane rating (Octane number) A measure of a fuel's resistance to self-ignition, hence a measure as well of the antiknock properties of the fuel.

Operating margin The operating margin refers to the changes in the operation of plants in an existing power system in response to a project-based activity (e.g., CDM).

Original Equipment Manufacturer (OEM) The original manufacturer of a vehicle or engine.

Oxides of Nitrogen (NO_x) Regulated air pollutants, primarily NO and NO_2 but including other substances in minute concentrations. Under the high pressure and temperature conditions in an engine, nitrogen and oxygen atoms in the air react to form various NO_x. Like hydrocarbons, NO_x are precursors to the formation of smog. They also contribute to the formation of acid rain.

Oxygenate A term used in the petroleum industry to denote fuel additives containing hydrogen, carbon and oxygen in their molecular structure. Includes ethers such as MTBE and ETBE and alcohols such as ethanol and methanol.

Oxygenated fuels Fuels blended with an additive, usually methyl tertiary butyl ether (MTBE) or ethanol to increase oxygen content, allowing more thorough combustion for reduced carbon monoxide emissions.

Oxygenated gasoline Gasoline containing an oxygenate such as ethanol or MTBE. The increased oxygen content promotes more complete combustion, thereby reducing tailpipe emissions of CO.

Ozone Tropospheric ozone (smog) is formed when volatile organic compounds (VOCs), oxygen, and NOx react in the presence of sunlight (not to be confused with stratospheric ozone, which is found in the upper atmosphere and protects the earth from the sun's ultraviolet rays). Though beneficial in the upper atmosphere, at ground level, ozone is a respiratory irritant and considered a pollutant.

Paraffins Group of saturated aliphatic hydrocarbons, including methane, ethane, propane and butane and noted by the suffix ane.

Particulate trap Diesel vehicle emission control device that traps and incinerates diesel particulate emissions after they are exhausted from the engine but before they are expelled into the atmosphere.

Peak watt (Wp) The peak watt is the unit by which the power output of PV modules is rated. It is defined as output under peak sunshine conditions of 25°C and irradiance of 1 kilowatt per square meter ($1 kW/m^2$ —roughly equivalent to the energy provided by the sun at noon in summer). Thus, if a 50Wp module is mounted in the tropics, at midday it should generate about 50W of electrical power; in practice, of course, the output will be slightly less, as the module will likely be operating at a temperature above 25°, which reduces conversion efficiency.

Petroleum fuel Gasoline and diesel fuel.

Permanence The longevity of a carbon pool and the stability of its stocks, given the management and disturbance environment in which it occurs.

Phase separation The phenomenon of a separation of a liquid or vapor into two or more physically distinct and mechanically separable portions or layers.

Photography, Audio-video production Training people to earn an income by taking pictures, recording video and sound.

Photovoltaics The use of lenses or mirrors to concentrate direct solar radiation onto small areas of solar cells, or the use of flat-plate photovoltaic modules using large arrays of solar cells to convert the sun's radiation into electricity.

Policies and measures Action by government to promote emissions reductions by businesses, individuals, and other groupings; measures include technologies, processes, and practices; policies include carbon or other energy taxes and standardized fuel-efficiency standards for automobiles.

Portable fueling system A system designed to deliver natural gas to fueling stations. Such systems are usually configured as tube trailers and are mobile. Fuel delivery usually occurs via over-the-road vehicles.

Practice An action or set of actions that affect the land, the stocks of pools associated with it or otherwise affect the exchange of greenhouse gases with the atmosphere.

Precision The repeatability of a measurement (e.g., the standard error of the sample mean).

Private fleet A fleet of vehicles owned by a non-government entity.

Propane (C_3H_8) A gas whose molecules are composed of three carbon and eight hydrogen atoms. Propane is often present in natural gas, and is also refined from crude petroleum. Propane contains about 18 kWh per cubic meter. In liquefied form called LPG.

Public Fueling Station Refers to fuelling station that is accessible to the general public.

Pump octane The octane as posted on retail gasoline dispensers as $(R+M)/2$; same as Antiknock Index.

Rainwater harvesting How to collect and store rain; the tools and techniques.

Reformulated gasoline (RFG) Gasolines that have had their compositions and/or characteristics altered to reduce vehicular emissions of pollutants.

Refueling emissions VOC vapors that escape from the vehicle fuel tank during refueling.

Regeneration The renewal of a stand of trees through either natural means (seeded on-site or adjacent stands or deposited by wind, birds, or animals) or artificial means (by planting seedlings or direct seeding).

Renewables Energy sources that, within a time frame that is brief relative to the earth's natural cycles, are sustainable; examples are non-carbon technologies such as solar energy, hydropower, waves and wind, as well as carbon-neutral technologies such as biomass.

Reid Vapor Pressure (RVP) A standard measurement of a liquid's vapor pressure in psi at 37.8 degrees Celsius. It is an indication of the propensity of the liquid to evaporate.

Research Octane Number (RON) The octane as tested in a single-cylinder octane test engine operated under less severe operating conditions. RON affects low to medium-speed knock and engine run-on. Research Octane is presented by the designation R in the $(R+M)/2$ equation and is the higher of the two numbers.

Retrofit To change a vehicle or engine after its original purchase, usually by adding equipment such as conversion systems.

Sanitation How to improve waste water treatment, make composting toilets, gray water systems; from small households to large villages.

Sequestration The process of removing and storing carbon dioxide from the atmosphere through, for example, land-use change, afforestation, reforestation, or enhancements of carbon in agricultural soils.

Shelter Improved building methods and materials, using cob, straw, adobe, passive solar design, etc.; the best of the new cutting-edge architectural breakthroughs.

Shifting agriculture A form of forest use common in tropic forests where an area of forest is cleared, or partially cleared, and used for cropping for a few years until the forest regenerates. Also known as "slash and burn agriculture," "moving agriculture," or "swidden agriculture."

Sink Any process or mechanism that removes a greenhouse gas from the atmosphere. A given pool (reservoir) can be a sink for atmospheric carbon if, during a given time interval, more carbon is flowing into it than is flowing out.

Sinks Places where CO_2 can be absorbed—the oceans, soil and detritus and land biota (trees and vegetation).

Smog A visible haze caused primarily by particulate matter and ozone. Ozone is formed by the reaction of hydrocarbons and NOx in the atmosphere.

Soil carbon pool Used here to refer to the relevant carbon in the soil. It includes various forms of soil organic carbon (humus) and inorganic soil carbon and charcoal. It excludes soil biomass (e.g., roots, bulbs, etc.) as well as the soil fauna (animals).

Solar cell The cell is the component of a photovoltaic system that converts light to electricity. Certain materials (silicon is the most used) produce a PV *effect* in reaction to sunlight; that is, sunlight frees electrons from sites within the materials. The solar cells are structured so that these freed electrons cannot return to their positively charged sites (or "holes") without flowing through an external circuit, thus generating a current. The cells are designed to absorb as much light as possible, but their individual outputs are small; hence, they are interconnected to provide power at an appropriate voltage.

Solar cooking How to use the sun to cook, make cookers for resale and/or micro-enterprise, from small household cookers to systems capable of cooking over 1000 meals per day; 2 weeks.

Solar electric training and demonstration Teaching people how to design, size, install, and maintain photovoltaic systems.

Solar greenhouses Growing food during cold months without a fuel source; passive solar greenhouse construction and use.

Solar pasteurizing Using the sun to purify water, create a mini-mfg./assembly business (cost EUR 30 for family of 4).

Solar space heating Solar heating methods that can be made or assembled locally.

Source Opposite of sink. A carbon pool (reservoir) can be a source of carbon to the atmosphere if less carbon is flowing into it than is flowing out of it.

Spark ignition engine Internal combustion engine, also called Otto motor, in which the charge is ignited electrically (e.g., with a spark plug).

Spill-over effect The economic effects of domestic or sectorial mitigation measures on other countries or sectors, which can be positive or negative and include effects on trade, carbon leakage, and the transfer and diffusion of environmentally sound technologies.

Stakeholders People or entities with interests that would be affected by a particular action or policy.

Stand See Forest Stand.

Standardized baseline emission rate A standardized (or multi-project) baseline emission rate can be calculated without reference to an individual project, based on a pre-defined methodology and characteristics of the regional power system.

Subsidy A direct payment from the government to an entity or a tax reduction to that entity, for implementing a practice the government wishes to encourage; greenhouse gas emissions can be discouraged by reducing fossil-fuel subsidies or granting subsidies for insulating buildings or planting trees.

Sulphur dioxide (SO$_2$) An EPA criteria pollutant.

Suspended particles Solid particles carried into the atmosphere with the gaseous products of combustion.

Tailpipe emissions EPA-regulated vehicle exhaust emissions released through the vehicle tailpipe. Tailpipe emissions do not include evaporative and refueling emissions.

Tax incentives In general, a means of employing the tax code to stimulate investment in or development of a socially desirable economic objective without direct expenditure from the budget of a given unit of government. Such incentives can take the form of tax exemptions or credits.

Technology transfer An exchange of knowledge, money, or goods that promotes the spread of technologies for adapting to or mitigating climate change; the term generally refers to the diffusion of technologies and technological co-operation across and within countries.

Tertiary Amyl Ethyl Ether (TAEE) An ether based on reactive C5 olefins and ethanol.

Tertiary Amyl Methyl Ether (TAME) An ether based on reactive C5 olefins and methanol.

Tetraethyl Lead or Lead An octane enhancer. One gram of lead increases the octane of one gallon of gasoline about 6 numbers.

Toluene Basic aromatic compound derived from petroleum and used to increase octane. The most common hydrocarbon purchased for use in increasing octane.

Toxic emission Any pollutant emitted from a source that can negatively affect human health or the environment.

Toxic substance A generic term referring to a harmful substance or group of substances. Typically, these substances are especially harmful to health. Technically, any compound that has the potential to produce adverse health effects is considered a toxic substance.

Transesterification A process in which organically derived oils or fats are combined with alcohol (ethanol or methanol) in the presence of a catalyst to form esters (ethyl or methyl ester).

Transportation Control Measures (TCM) Restrictions imposed by state or local governments to limit use or access by vehicles during certain times or subject to specific operating requirements, e.g., high-occupancy vehicle lanes.

Tropospheric oxone (O₃) Oxygen in condensation form in the lowest stratum of the atmosphere, otherwise known as smog.

Uptake The addition of carbon to a pool. A similar term is "sequestration."

Vapor pressure or volatility The tendency of a liquid to pass into the vapor state at a given temperature. With automotive fuels, volatility is determined by measuring RVP.

Variable Fuel Vehicle (VFV) A vehicle that has the capacity of burning any combination of gasoline and an alternative fuel. Also known as a flexible fuel vehicle.

Vehicle conversion Retrofitting a vehicle engine to run on an alternative fuel.

Volatile Organic Compound (VOC) Reactive gas released during combustion or evaporation of fuel. VOCs react with NOx in the presence of sunlight and form ozone.

Voluntary measures Measures to reduce greenhouse gas emissions that are adopted by firms or other actors in the absence of government mandates; they can involve making climate-friendly products or processes more readily available or encouraging to incorporate environmental values in their market choices.

Water pumping How to size and specify the correct pump, how to install and maintain systems.

Water pump manufacturing How to build and maintain "ram" water pumps that use the force of water falling to pump a percentage of the water uphill; making manually operated well pumps.

Watershed management Protecting, preserving, and restoring one of our most essential resources.

Wood Alcohol See Methanol.

Wood products Products derived from the harvested wood from a forest, including fuel wood and logs and the products derived from them such as sawn timber, plywood, wood pulp, paper, etc.

Xylene An aromatic hydrocarbon derived from petroleum and used to increase octane. Highly valued as a petrochemical feedstock. Xylene is highly photo chemically reactive and, as a constituent of tailpipe emissions, is a contributor to smog formation.

Zero Emission Vehicle (ZEV) A vehicle that emits no tailpipe exhaust emissions. ZEV credits can be banked within the Consolidated Metropolitan Statistical Area.

ABBREVIATIONS AND ACRONYMS

□	unknown or zero
~	approximately
<	less than
>	greater than
10^{12}	tera (T)
10^{15}	peta (P)
10^{18}	exa (E)
10^{3}	kilo (k)
10^{6}	mega (M)
10^{9}	giga (G)
AC	alternating current
API	American Petroleum Institute
B/d	barrels/day
bbl	barrel
bcm	billion cubic meters
billion	10^{9}
BOO	build, own, operate
BOT	build, operate, transfer
bscf	billion standard cubic feet
Btu	British thermal unit
BWE	Bundesverband Wind -Energy
CH_4	Methane
CHP	combined production of heat and power
CIS	Commonwealth of Independent States
cm	centimeter
CNG	compressed natural gas
CO	Carbon monoxide
CO_2	Carbon dioxide
Convention	United Nations Framework Convention on Climate Change
CUM or cum	Cubic meters
d	day
DC	direct current
DOWA	deep ocean water applications
ECE	Economic Commission for Europe
EJ	Exajoule
EJ	Exajoules
ETBE	ethyl tertiary butyl ether
EU	European Union
EUROSOLAR	European Association for Renewable Energies
FAO	UN Food and Agriculture Organization

FBC	Fluidized Bed Combustion
FC	fuel cell
FSU	former Soviet Union
GCV	gross calorific value
GDP	gross domestic product
GEF	Global Environment Facility
GHG	greenhouse gas
GJ	Gigajoule (10^9 joules)
GNP	gross national product
GOs	governmental organizations
GTCC	gas turbine/steam turbine combined cycle system
Gtoe	Giga tons of oil equivalent
GW	gigawatt (10^9 watts)
GW_e	gigawatt electricity
GWh	gigawatt hour
GWP	global warming potential
h	hour
ha	hectare
HC	hydrocarbon
HWR	heavy water reactor
Hz	hertz
IAEA	International Atomic Energy Agency
IBRD	International Bank for Reconstruction and Development
IEA	International Energy Agency
IIASA	International Institute for Applied Systems Analysis
IPCC	Intergovernmental Panel on Climate Change
IPP	independent power producer
IRENA	International Renewable Energy Agency
J	joule
kboe	thousand barrels of oil equivalent
kcal	kilocalorie
kg	kilogram
kgcr	kilogram of coal replacement
kgoe	kilogram of oil equivalent
kl	kilo liters (thousand liters)
km	kilometer
Km2	square kilometer
kt	thousand tons (10^3 metric tons)
ktoe	thousand tons (10^3 metric tons) of oil equivalent
Kw	kilowatt electricity
kWh	kilowatt-hour (10^3 watts per one hour)
kW_p	kilowatt peak
kW_t	kilowatt thermal
lb	pound (weight)
LNG	liquefied natural gas
LPG	liquefied petroleum gas
m	meter

m/s	meters per second
m^2	square meter
m^3	cubic meter
mb	Millibar
MJ	Megajoule
Ml	Megaliter
mm	millimeter
MNES	Ministry of Non-conventional Energy Sources (India)
mPa s	millipascal second
MPa	Megapascal
mt	million tons
MtC	million tons of carbon
MtCe	million tons of carbon equivalent
Mtce	million tons of coal equivalent
Mtcr	million tons of coal replacement
Mtoe	million tons of oil equivalent
mtoe	million tons of oil equivalent
MW	megawatts (10^6 watts)
MW_e	megawatt electricity
MWh	megawatt hour (10^6 watts x one hour)
MW_p	megawatt peak
MW_t	megawatt thermal
N	negligible
N_2O	Nitrous oxide
NCV	net calorific value
NEA	Nuclear Energy Agency
NGLs	natural gas liquids
NGOs	non-governmental organizations
Nm^3	normal cubic meter
NO_x	nitrogen oxides (or oxides of nitrogen)
NPP	net primary production
NPP	nuclear power plant
OAPEC	Organization of Arab Petroleum Exporting Countries
OECD	Organization for Economic Co-operation and Development
OLADE	Latin American Energy Organization
OPEC	Organization of the Petroleum Exporting Countries
OTEC	ocean thermal energy conversion
OWC	oscillating water column
p.a. or pa	Per annum
Parties	Parties to the UN Framework Convention on Climate Change
PJ	Petajoule (10^{15} joules)
ppm	parts per million
PPP	purchasing power parity
PV	photovoltaic
R&D	research and development
R, D&D	research, development and demonstration
R/P	reserves/production

RAP	Regional Office for Asia and the Pacific
RD&D	research, development and demonstration
rpm	revolutions per minute
SER	Survey of Energy Resources
SO_2	Sulphur dioxide
t	ton (metric ton)
tC	ton of carbon ($^{44}/12$ ton CO_2)
tce	ton of coal equivalent
tcf	trillion cubic feet
TFC	total final consumption of energy (IEA definition)
TJ	Terajoule
toe	ton of oil equivalent (10^7 kcal)
tpa	tons per annum
TPES	total primary energy supply (IEA definition)
trillion	10^{12}
ttoe	thousand tons of oil equivalent
tU	tons of uranium
TWh	Terawatt hour (10^{12} per one hour)
U	uranium
UN	United Nations
UNDP	United Nations Development Program
UNEP	United Nations Environment Program
UNFCCC	United Nations Framework Convention on Climate Change
VAT	value added tax
W	watt
WB	World Bank
WCRE	World Council for Renewable Energies
WEC	World Energy Council
WEO	World Energy Outlook
W_p	watts peak
WRI	World Resource Institute
wt	weight
yr	year

CONVERSION FACTORS

UNITS AND CONVERSIONS

General conversion factors for energy:

To:	TJ	Gcal	Mtoe	MBtu	GWh
From:			multiply by:		
TJ	1	238.8	2.388×10^{-5}	947.8	0.2778
Gcal	4.1868×10^{-3}	1	10^{-7}	3.968	1.163×10^{-3}
Mtoe	4.1868×10^{4}	10^{7}	1	3.968×10^{7}	11630
MBtu	1.0551×10^{-3}	0.252	2.52×10^{-8}	1	2.931×10^{-4}
GWh	3.6	860	8.6×10^{-5}	3412	1

Conversion factors for mass:

To:	kg	t	lt	st	lb
From:			multiply by:		
kilogram (kg)	1	0.001	9.84×10^{-4}	1.102×10^{-3}	2.2046
ton (t)	1000	1	0.984	1.1023	2204.6
long ton (lt)	1016	1.016	1	1.12	2240
short ton (st)	907.2	0.9072	0.893	1	2000
pound (lb)	0.454	4.54×10^{-4}	4.46×10^{-4}	5.0×10^{-4}	1

Conversion factors for volume:

To:	gal U.S.	gal U.K.	bbl	ft³	l	m³
From:			multiply by:			
U.S. gallon (gal)	1	0.8327	0.02381	0.1337	3.785	0.0038
U.K. gallon (gal)	1.201	1	0.02859	0.1605	4.546	0.0045
barrel (bbl)	42	34.97	1	5.615	159	0.159
cubic foot (ft³)	7.48	6.229	0.1781	1	28.3	0.0283
liter	0.2642	0.220	0.0063	0.0353	1	0.001
cubic meter (m³)	264.2	220.0	6.289	35.3147	1000	1

Decimal prefixes:

10^1	deca (da)	10^{-1}	deci (d)
10^2	hecto (h)	10^{-2}	centi (c)
10^3	kilo (k)	10^{-3}	milli (m)
10^6	mega (M)	10^{-6}	micro (μ)
10^9	giga (G)	10^{-9}	nana (n)
10^{12}	tera (T)	10^{-12}	pico (p)
10^{15}	peta (P)	10^{-15}	femto (f)
10^{18}	exa (E)	10^{-18}	atto (a)

Some of the conversion factors you may need to assess your site's feasibility:

1 cubic foot (cf)	7.48 gallons;
1 cubic foot per second (cfs)	448.8 gallons per minute (gpm);
1 inch	2.54 centimeters;
1 foot	0.3048 meters;
1 meter	3.28 feet;
1 cf	0.028 cubic meters (cm);
1 cm	35.3 cf;
1 gallon	3.785 liters;
1 cuff	28.31 liters;
1 cfs	1,698.7 liters per minute;
1 cubic meter per second (cm/s)	15,842 gpm;
1 pound per square inch (psi) of pressure	2.31 feet (head) of water;
1 pound (lb)	0.454 kilograms (kg);
1 kg	2.205 lbs.;
1 kilowatt (kW)	1.34 horsepower (hp);
1 hp	746 watts.

Energy Conversion and the Related WEC Conversions

In this Statement the conversion convention is the same as that used in ETW, namely that the generation of electricity from hydro (large and small scale), nuclear, and other new renewables (wind, solar, geothermal, oceanic but excluding modern biomass), has a theoretical efficiency of 38.46%. This convention, together with the use of the actual efficiencies (based on the low heating value) for plants using oil or oil products, natural gas or solid fuels (coal, lignite and biomass), guarantees a good comparability in terms of primary energy. However, for the record, WEC has now adopted in all of its recent publications the new conversion convention used by the IEA. New renewables and hydro are assumed to have a 100% efficiency conversion, except for geothermal (10% efficiency). For nuclear plants (excluding breeders) the theoretical efficiency is 33%. For the sake of continuity with ETW, these new conventions are <u>not</u> used in this Statement.

Conversion Factors and Energy Equivalents

1	calorie (cal)	= 4.18	J
1	joule (J)	= 0.239	cal
1000	KWh	= 3.6	GJ
1	ton of oil equivalent (net, low heat value)	= 42	GJ = 1 toe
1	ton of coal equivalent (standard, LHV)	= 29.3	GJ = 1 tce
1000	m^3 of natural gas (standard, LHV)	= 36	GJ
1	ton of natural gas liquids	= 46	GJ
1	toe	= 10 034	Mcal
1	tce	= 7000	Mcal
1000	m^3 of natural gas	= 8600	Mcal
1	ton of natural gas liquids	= 1000	Mcal
1	tce	= 0.697	toe
1000	m^3 of natural gas	= 0.857	toe
1	ton natural gas liquids	= 1.096	toe
1000	kWh	= 0.086	toe
1	ton of fuel wood	= 0.380	toe
1	barrel of oil	= 159	liters
1	barrel of oil	= Approx. 0.136	tons
1	cubic foot	= 0.0283	cubic meters

Because of rounding, some totals may not exactly equal the sum of their component parts, and some percentages may not agree exactly with those calculated from the rounded figures used in the tables.

Conversion Factors and Energy Equivalents
Basic Energy Units

1 joule (J)	= 0.2388 cal
1 calorie (cal)	= 4.1868 J
(1 British thermal unit [Btu]	= 1.055 kJ = 0.252 kcal)

WEC Standard Energy Units

1 ton of oil equivalent (toe)	= 42 GJ (net calorific value)	= 10 034 Mcal
1 ton of coal equivalent (tce)	= 29.3 GJ (net calorific value)	= 7 000 Mcal

Note: the ton of oil equivalent currently employed by the International Energy Agency and the United Nations Statistics Division is defined as 10^7 kilocalories, net calorific value (equivalent to 41.868 GJ)

Volumetric Equivalents

1 barrel	= 42 US gallons	= approx. 159 liters
1 cubic meter	= 35.315 cubic feet	= 6.2898 barrels

Electricity

1 kWh of electricity output	= 3.6 MJ	= approx. 860 kcal

Representative Average Conversion Factors

1 ton of crude oil	= approx. 7.3 barrels
1 ton of natural gas liquids	= 45 GJ (net calorific value)
1000 standard cubic meters of natural gas	= 36 GJ (net calorific value)
1 ton of uranium (light-water reactors, open cycle)	= 10 000 − 16 000 toe
1 ton of peat	= 0.2275 toe
1 ton of fuel wood	= 0.3215 toe
1 kWh (primary energy equivalent)	= 9.36 MJ = approx. 2 236 Mcal

Note: actual values vary by country and over time
Because of rounding, some totals may not agree exactly with the sum of their component parts

INVENTORY OF PV SYSTEMS FOR SUSTAINABLE RURAL DEVELOPMENT

Type of PV Application	Typical System Design	Existing Examples
Applications in the agricultural sector		
Lighting and cooling for poultry factory for extended lighting and increased production	50-150 Wp, electronics, battery, several TL-lights, fan	Egypt, India, Indonesia, Vietnam, Honduras
Irrigation	900 Wp, electronics, small DC or AC pump and water tank	India, Mexico, Chile
Electric fencing for grazing management	2-50 Wp panel, battery fence charger	USA, Australia, New Zealand, Mexico, Cuba
Pest control (moth)	Solar lanterns used to attract moths away from field	India (Winrock Intl.)
Cooling for fruit preservation	PV/wind hybrid systems or 300-700 Wp PV with DC refrigerators (up to 300 lt.)	Indonesia (Winrock Intl.)
Veterinary clinics	300 Wp, batteries, electronics refrigerator/ freezer, 2 TL-lights	Syria (FAO project)
Cattle watering	900 Wp, electronics DC/ AC pump, water reservoir	USA, Mexico, Australia
Aeration pumps for fish and shrimp farms	800 Wp, batteries (500 Ah), electronics, DC engine, paddle wheel, for 150m^2 pond	Israel, USA
Egg incubator	Panel up to 75 Wp, integrated box + heating element for hatching 60 eggs	India (Tata/BPSolar), Philippines (BIG-SOL project)
Crop spraying	5 Wp, sprayer	India (southern states), but cancelled from product package by BPSolar

Type of PV Application	Typical System Design	Existing Examples
Applications in cottage industry		
Tailor workshop	50-100 Wp system with DC lights and electric sewing machine	Several countries (i.e., NREL projects)
Electronic repair workshop	50-100 Wp for DC lights and soldering iron	Bangladesh (Grameen Shakti project) India, Indonesia
Gold jewelry workshop	60 Wp system with DC lights and soldering iron	Vietnam (SELF project)
Bicycle repair workshop	80 Wp system for DC lights and DC small drill	Conceptual: Vietnam — Ha Tinh Province (IFAD project)
Handicrafts workshop (small woodwork, bamboo, basket weaving, etc.)	60-100 Wp system for DC lights and DC small tools	Nepal, Vietnam
Trekking/eco-tourism lodges	Solar lanterns, SHSs and larger PV systems for lights and refrigeration	Nepal, India, Peru, Trinidad and Tobago
Pearl Farms	0.4 − 1 kW PV system to power craft workshops with drills, pumps, lights & compressor	Examples in French Polynesia (Solar energy)
Applications in the commercial service sector		
Village cinema	100-150 Wp system with DC lights and Color TV + VCR or satellite	Dominican Republic (ENERSOL project), Vietnam (Solarlab), Honduras
Battery charging stations	0.5 − 3 kWp systems with DC battery chargers for kWh sales to households and micro-enterprises	Morocco (Noor Web), Philippines (NEA), Senegal, Thailand, Vietnam (Solarlab), India, Bangladesh
Micro-utility	50 Wp, electronics, battery, 5 − 7 TL ("rented out")	India, Bangladesh (Grameen Shakti project)
Rent-out of solar lanterns for special occasions (weddings, parties, reunions)	Solar lanterns (5 − 10 Wp)	India (NEC) as part of a youth program

Type of PV Application	Typical System Design	Existing Examples
	Applications in cottage industry	

Applications in the commercial service sector (continued)

Lights, radio/TV and small appliances such as blenders for restaurants, shops and bars	20-300 Wp, electronics, battery, appliance, inverter (if necessary)	Many countries, incl. Karaoke bar in Philippines (NEA)
Trekking/eco-tourism lodges	Solar lanterns, SHSs and larger PV systems for lights and refrigeration	Nepal, India, Peru, Trinidad and Tobago, Mexico
Cellular telephone service	A 50 Wp system with 2 lights and a socket to charge cellular phone batteries	Bangladesh (Grameen Shakti project), Africa
Computer equipment in rural offices	$8 - 300$ Wp systems powering lights, fax, TV, etc.	Bangladesh, Costa Rica, Chile
Internet server for E-commerce	Integrated in multifunctional solar facility (> 1 kW)	West Bank (Greenstar project)

Applications for basic social services

Health clinics	150-200 Wp, electronics, deep-cycle batteries, small refrigerator/ freezer	Many countries (WHO standards)
Potable water pumping	1–4 kWp, electronics, pump, reservoir (generally no batteries needed)	Many countries, e.g. large project in Sahelian countries (EU-project)
Water purification	PV to power UV or ozone water purifiers (0.2-0.3 Wh/liter)	Many countries, e.g. China, Honduras, Mexico, West Bank
Water desalination	$1 - 2$ kWp needed to power reverse osmosis or other water desalination units for 1 m^3 per day	Italy, Japan, USA, Australia, Saudi United Arab Emirates
Internet server for telemedicine	Integrated in multifunctional solar facility (> 1 kW)	West Bank (Greenstar project)
Schools and Training centers	PV systems for powering lights, TV/VCR, PCs	Many countries: China, Honduras, Mexico, the Philippines, Mali
Street light	35/70 Wp, electronics, battery, 1 or 2 CFL	India, Indonesia, the Philippines, Brazil, Mali, Iraq

PROJECT "SOLARTECH SUD," SOLAR ECO-VILLAGE ZARZIS - DJERBA TUNISIA

The Djerba Zarzis eco-village project revolves around four main components as part of an integrated approach: the technological component, training, farming and eco-industrial component.

The construction of the first eco-solar village promoted by Solartech covers an area of 160 ha and an estimated cost between 130 million and 160 million Tunisian dinars.

The project's main objectives are:

- Make renewable energy, particularly solar energy, a lever for sustainable development of southeast Tunisia.
- Develop areas of expertise: renewable energy (particularly solar), desalination of water (especially seawater) and organic farming.
- Enhance synergies between education, research and production in the specialties of the village (renewable energy and organic farming).

This project is fully integrated within the framework of national priorities and is part of the Tunisian Solar Plan (TSP). It will support and promote the Tunisian policy in the field of energy management and sustainable development.

The south of Tunisia is very rich in vernacular architecture. In numerous examples of such architecture, we can witness many forms of sustainable development, through the harmony of the elements, their relevance to the actual climate and their roots in a long cultural tradition. The proposed

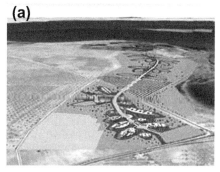

Figure A5.1a *Virtual image of the "Olive Branch."* (See color plate 51.)

Figure A5.1b *Solar Eco-Village Zarzis.* (See color plate 52.)

(C)

Figure A5.1c *Map of Tunisia. (Emna LATIRI, Architect 2012)*

Figure A5.2 *Virtual image of the Solar-Eco village. (Emna LATIRI, Architect 2012).* (See color plate 53.)

development seeks to reflect, in terms of images and spaces, simple forms enshrined in logic of a sustainable developmental oriented project. The architectural concept is inspired from an olive tree and expressed by a branch as a symbolic expression of conceptual choices of Solartech-Sud. The olive branch evokes the sustainable, the desirable, the natural, the adaptable and the scalable. Its natural growth is an element that generates dynamics while exploring the plan. The architectural image is linked to specific architectural references to the south of Tunisia, which may eventually be used to give authentic and local character to the image of the Solartech South project. The interior of the premises will be conforming to international standards in terms of features and amenities (Emna Latiri, 2012, architect, personal communication, www.solartech-sud.com/eng/).

SOLAR PARK VECHELDE (KRAFTFELD VECHELDE GMBH & CO. KG)

Individuals with innovative ideas and a spirit of implementation can move the world forward. One of them is pioneer Frank Ziegeler who converted an old sugar factory into a solar park for a sustainable future power supply. The heart of the project is a ground-mounted solar power system with 1.6 megawatts, one of the largest in South-East Lower Saxony. In on-going construction, up to 5,500 solar modules will be built in the future to fulfill the annual electricity needs of the towns of Wierthe and Sonnenberg.

Location: Wierthe Vechelde, Germany

Operator: Kraftfeld Vechelde GmbH & Co. KG

System power: 1.153.35 kWp

Annual Production: approx. 1,048,395 kWh (909 kWh/kWp)

CO_2 avoided: Approx. 733.9 tons/year

The International Research Centre for Renewable Energy (IFEED) has opened an office here that supports research and training activities.

Figure A6.1 *Aerial photo of the Solar Park Wierthe, Germany.* (See color plate 54.)

Figure A6.2 *A boy (Janosch Goosse) is expecting a sustainable and safe power supply for his future.* (See color plates 55 and 56.)

Figure A6.3 *(a) First solar desalination unit and (b) electric car are demonstrated at IFEED-Research Centre and Solar Park Vechelde.* (See color plates 57 and 58.)

SOLAR LAUNDRY, ETERNAL UNIVERSITY, BARU SAHIB, INDIA

The first private university under The Kalgidhar Trust in the interior Himalayan region at Baru Sahib was established in April 2008, presently imparting value-based education in 24 streams, viz. Arts, Divine music, Spiritual sciences, Medical, Engineering and other fields of higher education including Renewable Energies. Graduates of this unique educational system are excellent in academics and carry high moral values. They have love for

Figure A7.1 *Eternal University, Baru Sahib, India.* (See color plate 59.)

Figure A7.2 *The Valley of Baru Sahib, Eternal University India with solar heating and power supply systems.* (See color plate 60.)

Figure A7.3 *Solar devices on the buildings and valley slopes.* (See color plate 61.)

Figure A7.4 *Laundry building supplied with electricity (PV) and heated water (solar thermal).* (See color plate 62.)

Figure A7.5 *Laundry building.* (See color plate 63.)

humanity, compassion for the weak and commitment for selfless service for mankind. In a way, the graduates of EU work toward establishing permanent peace and universal brotherhood in the world as envisaged by the founding fathers.

MANUAL AND/OR SOLAR POWERED WATER TREATMENT SYSTEM

A filtration technology that makes it possible to eliminate bacteria and viruses in the water as well as removing all the dirt and at the same time preserving it's natural attributes. This allows most surface water sources to be utilized for drinking water! The machine can be operated by hand or electrical pumps that are powered by renewable energies. Depending on the effort, the way of pumping and the quality of the water, about 500 to 1000 liters of clean water per hour can be produced, enough to supply a whole village. The filter can be used for several years if looked after carefully. After filtration, the water has no need to be boiled, which saves money and energy. Moreover, there is less need for firewood, which preserves the trees and the environment.

On February 27, 2010, the first Afro-Prototype of the "Best Water Source" Water Processing Machine went on-line in Uganda, and in 2012 units will also be produced in Nigeria.

Figure A8.1 *(a) With external electrical pump; (b) Clean water from very dirty water.* (See color plate 64.)

REFERENCE

Hauseder, E.. Best Water Source. Austria. Available at: hauseder@ettl.net.

REFERENCES

Aitken, Donald W. Transitioning to a Renewable Energy Future, International Solar Energy Society, January 2010, p. 3.

Allen, C. 2011. German village achieves energy independence and then some [WWW] BioCycle. Available from http://www.infiniteunknown.nct/2011/08/22/german-village-wildpoldsried-pop-2600-produces-321-percent-more-energy-than-it-needs-and-is-generating-5-7-million-in-annual-revenue/

American Wind Energy Association (AWEA) 2011. 2010 U.S. Small Wind Turbine Market Report [WWW]. Available from http://www.awea.org/learnabout/smallwind/upload/2011_turbine_market_report.pdf

Andrews, David. 2006, Senior Technical Consultant, Biwater Energy. A talk originally given by as the Energy Manager at Wessex Water at an Open University Conference on Intermittency, 24 January 2006, National Grid's use of Emergency. Diesel Standby Generator's in dealing with grid intermittency and variability. Potential Contribution in assisting renewables Ashden Awards. "Micro-hydro". Retrieved 2009-06-29.

BBC News 2008. "Airline in first biofuel flight" [WWW] BBC News. Available from http://news.bbc.co.uk/2/hi/7261214.stm

Bertani, Ruggero; Thain, Ian (July 2002), "Geothermal Power Generating Plant CO2 Emission Survey", IGA News (International Geothermal Association) (49): 1–3, retrieved 2009-05-13.

Best Water Source, Ernst Hauseder – Austria, hauseder@ettl.net.

Biopact Team 2008, "Report: Biogas can replace all EU natural gas imports" [WWW] Biopact, Available from Bloomberg New Energy Finance, 2011, "Global New Investments in Renewable Energy" UNEP SEFI, Frankfurt School, Global Trends in Renewable Energy Investment. Bloomberg, L P 2012. New Energy Finance [WWW]. Available from http://www.newenergyfinance.com/PressReleases/view/186

"BTM Forecasts 340-GW of Wind Energy by 2013". Renewableenergyworld.com. 27 March 2009.

Boston Consulting Group 2009. "Electricity Storage, Making Large-Scale Adoption of Wind and Solar Energies a Reality", http://www.bcg.no/expertise_impact/Industries/Energy_Environment/PublicationDetails.aspx?id=tcm:106-41977

"BP Statistical World Energy Review 2011" (XLS). Retrieved 8 August 2011.

BSU Solar 2007, German Solar Industry Association

Business Dictionary 2012, "Solar energy" [WWW]. Available from http://www.businessdictionary.com/definition/solar-energy.html#ixzz1lsoyXA8S

Carbon Trust, Future Marine Energy. Results of the Marine Energy Challenge: Cost competitiveness and growth of wave and tidal stream energy, January 2006.

Clean Edge (2009). Clean Energy Trends 2009, pp. 1-4.

Clean energy projected growth 2007-2017. Clean Edge (2008) Energy Trends Report (GGByte at en.wikipedia 2012), http://en.wikipedia.org/wiki/File:Re_investment_2007-2017.jpg

Cothran, Helen. Energy Alternatives., Greenhaven Press, March 1, 2002.

Community Energy Scotland 2011, *Community renewable energy toolkit* [WWW] Available from http://www.scotland.gov.uk/Resource/Doc/264789/0079289.pdf

Danish Energy Agency, Heat supply: Goals and means over the years, http://dbdh.dk/images/uploads/pdf-diverse/varmeforsyning%20i%20DK%20p%C3%A5%20engelsk.pdf Department of Energy (DOE) Office of Energy Efficiency and Renewable Energy (EERE), https://www.eere-pmc.energy.gov/PMC_News/EERE_Program_News_3-08.aspx

DiChristina, Mariette (May 1995). "Sea Power". Popular Science: 70–73. Retrieved November 2011.

Doughty, Daniel, Dudei, H. Paul C., Akhil, Abbas A., Clark, Nancy, H., and Boyes, John D. Batteries for Large-Scale Stationary Electrical Energy Storage, The Electrochemical Society Interface 2010, http://www.electrochem.org/dl/interface/fal/fal10/fal10_p049-053.pdf

Duke Energy 2012, Generating Electricity with Pumped-Storage Hydro, http://www.duke-energy.com/about-energy/generating-electricity/pumped-storage-faq.asp

Electrical Storage 2012, http://www.microchap.info/electrical_storage.htm

Emna LATIRI, Architect, Tunisia, Personal communication 2012, www.solartech-sud.com/eng/

Energy Independence and Security Act, U.S.C. Title XIII Sec. 1301 (2007). Retrieved from http://energy.gov/sites/prod/files/oeprod/DocumentsandMedia/EISA_Title_XIII_Smart_Grid.pdf

Eternal University, 2011, Baru Sahib, Via Rajgarh, Sirmour, Himachal Pradesh, India – 173101.

European Commission 2006. *European SmartGrids Technology Platform: Vision and Strategy for Europe's Electricity Networks of the Future*, Luxembourg, EUR 22040. Federation of German Industries/BDI Initiative Internet of Energy 2010. *Internet of Energy ICT for Energy Markets of the Future: The Energy Industry on the Way to the Internet Age* (translation of the brochure "Internet der Energie – IKT für Energiemärkte der Zukunft" published in Germany, December 2008, to which information about the German government's E-Energy model projects has been added). Available from http://www.bdi.eu/bdi_english/download_content/Marketing/Brochure_Internet_of_Energy.pdf

Gipe, P. 2012. From data by Unendlich viel energie, *Windworks*, January 5, 2012 [WWW]. Available from http://wind-works.org/coopwind/CitizenPowerConferenceto beheldinHistoricChamber.html

Gipe, P. 2012. "Germany renewable energy generation is owned by its own citizens" [WWW], January 5, 2012, *Windworks*, http://wind-works.org/coopwind/CitizenPowerConferencetobeheldinHistoricChamber.html

Glassley, William E. Geothermal Energy: Renewable Energy and the Environment CRC Press, 2010.

Gloystein, Henning (November 23, 2011). "Renewable energy becoming cost competitive, IEA says". Reuters.

Geothermal Energy Association. Geothermal Energy: International Market Update May 2010, pp. 4-6, 7.

Gore, Al. (2009). *Our Choice*, p. 58, Bloomsbury Publishing PLC, ISBN: 0747590990.

Eberle, Ulrich; von Helmolt, Rittmar (2010-05-14). "Sustainable transportation based on electric vehicle concepts: a brief overview". Royal Society of Chemistry, http://pubs.rsc.org/en/Content/ArticleLanding/2010/EE/c001674h

El Bassam, N. & Maegaard, P. 2004. Integrated renewable energy for rural communities: Planning guidelines, technologies and applications, Elsevier Science, Radarweg 29, Amsterdam 1043 NX.

El Bassam, N. (2010). Handbook of bioenergy crops: A complete reference to species, development and applications, 2nd ed., Routledge, Taylor & Francis Group Ltd, Oxford.

El Bassam, N. (Ed.) 1998. Energy plant species: Their use and impact on environment and development, Routledge, Taylor & Francis, Inc., Oxford.

El Bassam, N. 1996. Renewable energy: Potential energy crops for Europe and the Mediterranean region, REU Technical Series 46, Food and Agriculture Organization of the United Nations (FAO), 200 S, Rome.

El Bassam, N. 1998a. "Biological Life Support Systems under Controlled Environments" in N El Bassam, et al. (eds.), Sustainable agriculture for food, energy and industry, James & James Science Publishers, Volume 2, London.

El Bassam, N. 1998b. Energy plant species: Their use and impact on environment and development, James & James Science Publishers, London.

El Bassam, N. 1999. Integrated Energy Farm Feasibility Study, SREN-FAO.

El Bassam, N. 2001. "Renewable Energy for Rural Communities," Renewable Energy, vol. 24, pp. 401–408

El Bassam, N. 2004. "Integrated Renewable Energy Farms for Sustainable Development in Rural Communities," In Biomass and agriculture, OECD, Paris, pp. 262–276. Available at www.oecd.org/agr/env

El Bassam, N. 2010. Handbook of bioenergy crops: A complete reference to species, development and applications, Routledge, Taylor & Francis Group Ltd, Oxford.

El Bassam, N. © 2009. Solar Oases: Transformation of Deserts into Gardens, www.ifeed. org.

Energy Independence and Security Act, U.S.C. Title XIII Sec. 1301 (2007). Retrieved from http://energy.gov/sites/prod/files/oeprod/DocumentsandMedia/EISA_Title_XIII_ Smart_Grid.pdf

Energy4You. 2012. Solar hot air [WWW]. Available from http://www.energy4you.net/ hotair.htm

Energy Society of Canada Inc. and Solar, October 21-24, 2000.

Etcheverry, J. 2008. Challenges and opportunities for implementing sustainable energy strategies in coastal communities of Baja California Sur, Mexico, University of Toronto.

European Commission 2006. European SmartGrids Technology Platform: Vision and Strategy for Europe's Electricity Networks of the Future, Luxembourg, EUR 22040.

European Photovoltaic Industry Association (2012). "Market Report 2011".

Exploit Nature-Renewable Energy Technologies by Gurmit Singh, Aditya Books, http://www.awsopenwind.org/downloads/documentation/ModelingUncertainty Public.pdf

Federal Ministry of Food, Agriculture and Consumer Protection, Berlin Office 11055 Berlin [WWW]. http://www.wege-zum-bioenergiedorf.de/metanavigation/datenschutz/

Federation of German Industries/BDI Initiative Internet of Energy 2010. Internet of Energy ICT for Energy Markets of the Future: The Energy Industry on the Way to the Internet Age (translation of the brochure "Internet der Energie – IKT für Energiemärkte der Zukunft" published in Germany, December 2008, to which information about the German government's E-Energy model projects has been added). Available from http:// www.bdi.eu/bdi_english/download_content/Marketing/Brochure_Internet_of_ Energy.pdf

Fridleifsson, Ingvar B.; Bertani, Ruggero; Huenges, Ernst; Lund, John W.; Ragnarsson, Arni; Rybach, Ladislaus (2008-02-11), O. Hohmeyer and T. Trittin, ed., The possible role and contribution of geothermal energy to the mitigation of climate change, Luebeck, Germany, pp. 59–80, retrieved 2009-04-06.

Fthenakis, V.; Kim, H. C. (2009). "Land use and electricity generation: A life-cycle analysis". Renewable and Sustainable Energy Reviews 13 (6–7): 1465.

Fuel Cells 2000 (2008). "Fuel Cell Basics.", http://www.fuelcells.org/basics/types.html

Gangwar, R. 2009. "Building community resilience towards climate change adaptation through awareness and education", paper presented at the seminar Energy and Climate in Cold Regions of Asia, 21–24 April, 2009, available at http://india.geres.eu/ docs/Seminar_proceedings/_3 Climate_Change_Impacts_and_Adaptation/Building% 20Community%20Resilience%20towards%20Climate%20Change%20Adaptation% 20through%20Awareness%20&%20Education.pdf

Gipe, P. 2012. "Germany renewable energy generation is owned by its own citizens" [WWW], January 5, 2012, Windworks, http://wind-works.org/coopwind/Citizen PowerConferencetobeheldinHistoricChamber.html

Gipe, P. 2012. From data by Unendlich viel energie, Windworks, January 5, 2012 [WWW]. Available from http://wind-works.org/coopwind/CitizenPowerConferenceto beheldinHistoricChamber.html

Global Energy Wind Council (GWEC), Press release Brussels, February 2, 2007, Global wind energy markets continue to boom – 2006 another record year (PDF).

Gorlov, A.M. Development of the helical reaction hydraulic turbine. Final Technical Report, The US Department of Energy, August 1998, The Department of Energy's (DOE) Information Bridge: DOE Scientific and Technical Information.

Hau, Erich. Wind turbines: fundamentals, technologies, application, economics Birkhäuser, 2006 ISBN 3-540-24240-6.

Hot Water Now [n.d.] Hot Water Adelaide, Solar, Gas & Heat Pump [WWW]. Available from www.hotwaternow.com.au/colchester_solar_panel.jpg

How Stuff Works, "Ocean currents". p. 4. Retrieved 2 November 2010, http://science. howstuffworks.com/environmental/earth/oceanography/ocean-current4.htm

Howtopedia 2010. Biomass (Technical Brief) [Online]. Available from http://en. howtopedia.org/wiki/Biomass_%28Technical_Brief%29

IEA Renewable Energy Working Party (2002). Renewable Energy… into the mainstream, p. 9.

Insource/Outsource: 2007-09-16, http://insourceoutsource.blogspot.com/2007_09_16_ archive.html

http://www.vlh-turbine.com/EN/html/History.htm

International Climate and Environmental Change Assessment Project (ICECAP) 2007, "Tidal power: Clean, reliable and renewable," accessed at http://icecap.us/images/ uploads/TIDAL_POWER.pdf

International Energy Agency (IEA) 2008. Summary Document Ad-Hoc Group on Science & Energy Meeting Scientific Breakthroughs for a Clean Energy Future 6-7 May, 2008 IEA Secretariat, Paris, France

International Energy Agency Recharge 2012. "China continues offshore push with 300MW wind farm plan" [WWW]. Available from http://www.rechargenews.com/energy/ wind/article297985.ece

Iraq Dream Homes, LLC (IDH) 2010 [WWW]. Available from http://www.iraq-homes. com/about_us-en.html

IZNE 2005. The Bioenergy Village Self-sufficient Heating and Electricity Supply Using Biomass, www.bioenergiedorf.de

IZNE 2007. Wärme- und Stromversorgung durch heimische Biomasse April 2007, Interdisziplinäres Zentrum für Nachhaltige Entwicklung (IZNE) der Universität Göttingen. - Projektgruppe Bioenergiedörfer, http://www.gar-bw.de/fileadmin/gar/ pdf/Energie_und_Klima/Juehnde.pdf

Johnson, Keith. Wind Shear: GE Wins, Vestas Loses in Wind-Power Market Race, Wall Street Journal, March 25 2009, accessed on January 7 2010.

Kaplan, S. M. (2009). Smart Grid. Electrical Power Transmission: Background and Policy Issues. The Capital.Net, Government Series. pp. 1-42.

Khan, M. Ali. (2007) (pdf). The Geysers Geothermal Field, an Injection Success Story, Annual Forum of the Groundwater Protection Council, retrieved 2010-01-25.

Korea Smart Grid Institute (KSGI). 2010, Korea's Jeju Smart Grid. [WWW]. Available from http://www.smartgrid.or.kr/10eng3-3.php.

Kraftfeld Vechelde GmbH & Co. KG, Fabrikstraße 6 38159 Vechelde, Germany.

Kroldrup, Lars. Gains in Global Wind Capacity Reported Green Inc., February 15, 2010.

Lamb, T. 2011. Driving ocean energy innovation in Scotland. Renewable Energy World. Viewed June 30, 2012 at http://www.renewableenergyworld.com/rea/news/article/2011/06/driving-ocean-energy-innovation-in-scotland

Leone, Steve (25 August 2011). "U.N. Secretary-General: Renewables Can End Energy Poverty". Renewable Energy World.

Levant Power Corp 2011. "Revolutionary GenShock Technology", http://www.levantpower.com/technology.html

Lund, H. 2010. Renewable energy systems: The choice and modeling of 100% renewable solutions, Academic Press.

Maegaard, P. 2008. "Denmark: Wind Leader in Stand-By", International Sustainable Energy Review.

Maegaard, P. 2009. Danish Renewable Energy Policy, article, WCRE*org homepage, September 2009.

Maegaard, Preben. Integrated Systems to Reduce Global Warming 2011, Springer Science+Business Media, LLC, 22. June 2011 (PDF).

Maegaard, P. 2009/2010. "Thisted: 100% Renewable Energy municipality." PowerPoint presentation.

Maegaard, P. 2010. "Wind Energy Development and Application Prospects, of Non-Grid-Connected Wind Power," in Proceedings of 2009 World Non-Grid-Connected Wind Power and Energy Conference, IEEE Press.

Maegaard, P. 2004. Wind energy development and application prospects of non-grid-connected wind power 2004, World Wind Energy Institute, World Renewable Energy Committee, Nordic Folkecenter for Renewable Energy, Denmark.

Mancini, Tom. 2006, "Advantages of Using Molten Salt". Sandia National Laboratories. Archived from the original on 2011-07-14, http://www.webcitation.org/60AE7heEZ

Månedlig elforsyningsstatistik, HTML-spreadsheet summary tab B58-B72 Danish Energy Agency, 18 January 2012.

Marloff, Richard H. 1978, Stresses in turbine-blade tenons subjected to bending, Experimental Mechanics, Volume 18, Number 1 - SpringerLink (PDF)

Marshall, M. 2012. New Scientist, issue 2850, accessed from http://www.newscientist.com/article/mg21328505.000-indias-panel-price-crash-could-spark-solar-revolution.html

Micro Hydro in the fight against poverty, http://tve.org/ho/doc.cfm?aid=1636&lang=English

Miller, Alasdair and Lumby, Ben, Utility Scale solar Power Plants, (PDF) 2011, written for the International Finance Corporation (IFC) and was implemented in partnership with The Global Environment Facility (GEF) and the Austrian Ministry of Finance.

Modular wind energy device - Brill, Bruce I. 2002, http://www.freepatentsonline.com/6481957.html

"Monthly Statistics – SEN". February 2012.

Morales, Alex (February 07, 2012). "Wind Power Market Rose to 41 Gigawatts in 2011, Led by China". Bloomberg; LP, http://www.bloomberg.com/news/2012-02-07/wind-power-market-rose-6-percent-to-41-gigawatts-led-by-china.html

NASA Science/Science News 2002. "How do photovoltaics work?" [WWW]. Available from http://science.nasa.gov/science-news/science-at-nasa/2002/solarcells/

NASA Science/Science News 2011, "The edge of sunshine" [WWW]. Available from http://science.nasa.gov/science-news/science-at-nasa/2002/08jan_sunshine/

National Renewable Energy Laboratory (2006). Nontechnical Barriers to Solar Energy Use: Review of Recent Literature, Technical Report, NREL/TP-520-40116, September.

Nemzer, J. 2012. "Geothermal heating and cooling", Cambridge, England, UK: Cambridge University Press, pp. 136–137, ISBN 978-0-521-66624-4.

New Albany Innovation Exchange, Ohio's First Electric Car Charging Station. Posted on September 15, 2011, http://www.innovatenewalbany.org/business/ohio%E2%80%99s-first-electric-car-charging station/

Nguyen, T.C. 2011. "Electric cars can now earn money for owners," Smart planet, September 29, 2011. Available from http://www.smartplanet.com/blog/thinking-tech/electric-cars-can-now-earn-money-for-owners/8756

Nitsch, F. 2007. BMU documentation

Northern Territory Government Australia: Roadmap to Renewable and Low Emission Energy in Remote Communities Report, http://www.greeningnt.nt.gov.au/climate/docs/Renewable_Energy_Report_FA.pdf

Ocean Energy Glossary 2007. Wave Energy Centre supported by the Co-ordinated Action of Ocean Energy EU funded Project (CA-OE) and with the Implementing Agreement on Ocean Energy Systems (IEA-OES) (PDF).

Ocean Thermal Energy Conversion 2011. Mineral Extraction, http://www.nrel.gov/otec/mineral_extraction.html

Offshore Wind 2012. "Atlantis to install tidal power farm in Gujarat, India" [WWW]. Available from http://www.offshorewind.biz/2012/02/01/atlantis-to-install-tidal-power-farm-in-gujarat-india/

Our Energy 2012. Geothermal energy facts. [WWW] Our Energy. Available from http://www.our-energy.com/energy_facts/geothermal_energy_facts.html

Pacific Gas and Electric Company 2007. First vehicle-to-grid demonstration, http://seekingalpha.com/article/31992-pacific-gas-and-electric-demonstrates-vehicle-to-grid-technology

Pacific Gas & Electric (PG&E) 2011. "Smart Grid Definition," Smart grid deployment plan. Available from http://www.neuralenergy.info/2009/09/pg-e.html

Peak Oil news & message boards 2011. "Exploring Hydrocarbon Depletion" [online]. Available at http://peakoil.com/what-is-peak-oil/

Perlin, John. (1999). From Space to Earth (The Story of Solar Electricity). Harvard University Press. ISBN 0-674-01013-2.

Pockley, Simon. Compressed Air Energy Storage (CAES), Prepared for Intro. to Renewable Energy 19/05/2008 (PDF).

PV Resources.com (2011). World's largest photovoltaic power plants.

QuantumSphere Inc. 2006. "Highly Efficient Hydrogen Generation via Water Electrolysis Using Nanometal Electrodes", http://www.qsinano.com/white_papers/2006_09_15.pdf.

Renewables in global energy supply: An IEA facts sheet (PDF) OECD, International Energy Agency (IEA) and Organization for Economic Cooperation and Development (OECD) 2007, 34 pages.

"Renewables Investment Breaks Records". Renewable Energy World. 29 August 2011.

Renewables Global Status Report 2006 Update, REN21, published 2006.

REN21 Renewables 2007. Global Status Report, 2008 Deutsche Gesellschaft für Technische Zusammenarbeit (GTZ) GmbH. www.ren21.net

REN21 (2010). Renewables 2010 Global Status Report pp. 9, 15, 26, 34, 53.

REN21 (2011). "Renewables 2011: Global Status Report". p. 11.

REN21 (2012). Renewables Global Status Report 2012, p. 17.

Renewable and Sustainable Energy Reviews 13 (6–7): 1465.

"Renewables". eirgrid.com. Retrieved 22 November 2010.

Renewable Communities 2008. "Renewable communities: Moving towards community resiliency" [online]. Available at http://renewablecommunities.wordpress.com

Renewable Energy Policy Network for the 21st Century (REN21) 2006–2011. "Renewables Global Status Report" [WWW]. Available from http://www.ren21.net/REN21Activities/Publications/GlobalStatusReport/tabid/5434/Default.aspx

ReNews Europe 2012. OE wins Wave Hub deal [WWW]. Available from http://renews. biz/story.php?page_id=71&news_id=1372

Research Institute for Sustainable Energy, "Microhydro". Retrieved 9 December 2010.

Rise & Shine. 2000. the 26th Annual Conference of the Solar, Halifax, Nova Scotia, Canada.

Ruby, J. U. 2006. Continuous headwind – pioneering the transition from fossil fuels and atomic energy to the renewable energies, Hovedland (Danish).

Ruwisch, V. B. Sauer. 2007 Bioenergy Village Jühnde: Experiences in rural self-sufficiency.I, IZNE, www.bioenergiedorf.de

Rybach, Ladislaus (September 2007). "Geothermal Sustainability", Geo-Heat Centre Quarterly Bulletin (Klamath Falls, Oregon: Oregon Institute of Technology) 28 (3): 2–7, ISSN 0276-1084, retrieved 2009-05-09.

Scheer, H. 2006. Energy autonomy: The economic, social and technological case for renewable energy, Earthscan, 2006.

Shahan, Z. 2011. "Small wind turbine market growing strong in U.S." [WWW]. Clean Technica. Available from http://cleantechnica.com/2011/09/25/small-wind-turbine-market-growing-strong-in-u-s/

Shankleman, Jessica. "The Queen's hydro energy scheme slots into place". BusinessGreen, 21 December 2011. Retrieved 21 July 2012.

Shenzhen JCN New Energy Technology CO, Ltd http://www.jcnsolar.com/help/help-0001,0002,0015.shtml

Silicon Laboratories 2011. The energy harvesting tipping point for wireless sensor applications [WWW]. White paper available from http://www.eetimes.com/electrical-engineers/education-training/tech-papers/4217176/The-Energy-Harvesting-Tipping-Point-for-Wireless-Sensor-Applications

Sills, Ben (August 29, 2011). "Solar May Produce Most of World's Power by 2060, IEA Says". Bloomberg LP, http://www.bloomberg.com/news/2011-08-29/solar-may-produce-most-of-world-s-power-by-2060-iea-says.html

"Small Photovoltaic Arrays". Research Institute for Sustainable Energy (RISE), Murdoch University. Retrieved 5 February 2010. (Modified by El Bassam).

Small Wind, U.S. Department of Energy National Renewable Energy Laboratory website. http://www.nrel.gov/wind/smallwind/

Smeets, E. Faaj, A., & Lewandowski, I. 2006. "Progress in Energy and Combustion Science," in A quick scan of global bioenergy potentials to 2050.

Solar Millennium AG 2011, http://www.solarmillennium.de/english/technology/references_and_projects/andasol-spain/index.html

Solar: photovoltaic: Lighting Up The World. Retrieved 19 May 2009.

Solar Power 50% Cheaper By Year End - Analysis Reuters, November 24, 2009.

Solar Tribune 2012. Guide to solar thermal energy [WWW]. Available from http://solartribune.com/solar-thermal-power/#.T0_PvfU0iuI

SolarWorld Group 2008. "Solar Power for the Vatican: Inauguration of the First Solar Power Plant for the Papal State," Bonn, [WWW]. Available from http://solarworld-usa.com/news-and-resources/news/vatican-inauguration-solar-power-for-papal-state.aspx

Stekli, J. 2010. "DOE CSP R&D: Storage Award Overview." U.S. Department of Energy: Energy Efficiency & Renewable Energy: Solar Energy Technologies [WWW]. Available from DOE CSP R&D: Storage Award Overview, DOE HQ | April 28, 2010.

Strategy Workshop, June 2011. "Green Settlements in Sub Saharan Africa: Future Land Use and Pathways to Wealth Creation" Loccum Protestant Academy, Germany.

Technical Specs of Common Wind Turbine Models 2007 [AWEO.org].

Tennessee Valley Authority (TVA) 2004. Energy [WWW]. Available from http://www.tva.gov/power/pumpstorart.htm "the Spanish electricity system: preliminary report 2011". January 2012. p. 13.

The Ashden Awards for Sustainable Energy, "Micro-hydro". Retrieved 20 November 2010.

The Free Dictionary 2002. McGraw-Hill Concise Encyclopedia of Engineering Energy storage, http://encyclopedia2.thefreedictionary.com/Energy+storage

The Energy Blog (2008). Largest tidal stream system installed. Weblog [Online] 06 April. Available from http://thefraserdomain.typepad.com/energy/ocean_power/

The Poul la Cour Foundation 2009. Wind power, the Danish way, from Poul la Cour to modern wind turbines. Self-Published, Denmark.

TheFreeDictionary.com. 2012. "Solar energy" [WWW]. Available from http://encyclopedia.thefreedictionary.com/solar+energy

Toke, D. 2007. "Evaluation and Recommendations" (DESIRE project – Dissemination strategy on Electricity balancing large scale integration of Renewable Energy).

"Trends in Photovoltaic Applications Survey report of selected IEA countries between 1992 and 2009, IEA-PVPS". Retrieved 8 November 2011.

Trieb, Dr. Franz. 2005. MED-CSP, Final Reports, German Aerospace Center (DLR) Institute of Technical Thermodynamics, Section Systems Analysis and Technology Assessment, Study commissioned by: Federal Ministry for the Environment, Nature Conservation and Nuclear Safety, Germany, Institute of Technical Thermodynamics, Section Systems Analysis and Technology Assessment, www.dlr.de/tt/med-csp

Trieb, Dr. Franz. 2006. TRANS-CSP.Final Reports, German Aerospace Center (DLR)

Institute of Technical Thermodynamics, Section Systems Analysis and Technology Assessment, Study commissioned by: Federal Ministry for the Environment, Nature Conservation and Nuclear Safety, Germany, Institute of Technical Thermodynamics, Section Systems Analysis and Technology Assessment, www.dlr.de/tt/trans-csp

Trieb, Dr. Franz. 2007. AQUA-CSP, Final Reports, German Aerospace Center (DLR)

Institute of Technical Thermodynamics, Section Systems Analysis and Technology Assessment, Study commissioned by: Federal Ministry for the Environment, Nature Conservation and Nuclear Safety, Germany, Institute of Technical Thermodynamics, Section Systems Analysis and Technology Assessment, http://www.dlr.de/tt/aqua-csp

Turcotte, D. L.; Schubert, G. (2002). "4", Geodynamics (2 ed.), Cambridge, England, UK: Cambridge University Press, pp. 136–137, ISBN 978-0-521-66624-4.

Union of Concerned Scientists, Clean Energy, How Geothermal Energy Works 2009, http://www.ucsusa.org/clean_energy/technology_and_impacts/energy_technologies/how-geothermal-energy-works.html

United Nations Environment Programme and New Energy Finance Ltd. (2007). Global Trends in Sustainable Energy Investment 2007: Analysis of Trends and Issues in the Financing of Renewable Energy and Energy Efficiency in OECD and Developing Countries (PDF) p. 3.

U.S. Department of Energy (DOE) 2001. Concentrating solar power commercial application study: Reducing water consumption of concentrating solar power electricity generation. A report to Congress. Available from http://www1.eere.energy.gov/solar/pdfs/csp_water_study.pdf

U.S. Department of Energy (DOE) 2011a. "Microhydropower System Components" [WWW], Energy Savers, Available from http://www.energysavers.gov/your_home/electricity/index.cfm/mytopic=11100

U.S. Department of Energy (DOE) 2011b. "How a Microhydropower System Works" [WWW], Energy Savers. Available from http://www.energysavers.gov/your_home/electricity/index.cfm/mytopic=11060

U.S. Dept. of Energy (DOE). "Gemasolar Thermosolar Plant". Concentrating Solar Power Projects. National Renewable Energy Laboratory (NREL), 24 October 2011.

U.S. Department of Energy (DOE) 2012. Geothermal technologies program. [WWW] Energy Efficiency & Renewable Energy. Available from http://www1.eere.energy. gov/geothermal/powerplants.html

U.S. Department of Energy (DOE) EERE 2011. "Photovoltaics" [WWW]. Available from http://www1.eere.energy.gov/solar/pdfs/52481.pdf

U.S. Department of Energy (DOE) EERE 2011a. Ocean Wave Power, accessed from http://www.energysavers.gov/renewable_energy/ocean/index.cfm/mytopic=50009

U.S. Department of Energy (DOE) EERE 2011b. Ocean Thermal Energy Conversion, accessed from http://www.eere.energy.gov/basics/renewable_energy/ocean_thermal_ energy_conv.html

U.S. Dept. of Agriculture, Forest Service 1961. Charcoal Production, Marketing and Use, #2213. Available from www.fpl.fs.fed.us/documnts/fplr/fplr2213.pdf

U.S. Department of Energy. "Smart Grid / Department of Energy". Retrieved 2012-06-18.

Union of Concerned Scientists 2009. How geothermal energy works. [WWW] Clean Energy. Available from http://www.ucsusa.org/clean_energy/technology_and_ impacts/energy_technologies/how-geothermal-energy-works.html

Varnon, Rob. Derecktor converting boat into hybrid passenger ferry, Connecticut Post website, December 2, 2010.

Vertical-Axis Wind Turbines | Symscape, http://www.symscape.com/blog/vertical_axis_ wind_turbine

Wagner, Leonard. 2008. Nanotechnology in the clean tech sector, Research report January 2008, http://www.moraassociates.com/publications/0801%20Nanotechnology.pdf

Wagner, Leonard. Overview of energy storage methods, Research report December 2007, http://www.moraassociates.com/publications/0712%20Energy%20storage.pdf

Wang, Ucilia (6 June 2011). "The Rise of Concentrating Solar Thermal Power". Renewable Energy World.

WEC Global Issues Map 2011, www.worldenergy.org/wec_news/press_releases/3264.asp

Whitlock, Charles H,. Brown, Donald E., Chandler, William S. and DiPasquale, Roberta C. Release 3 NASA Surface Meteorology and Solar Energy Data Set for Renewable Energy Industry Use. 1993, collaborations: Natural Resources Canada, National Renewable Energy Laboratory, Solar Energy International, Sun Frost, Inc., the Center for Renewable Energy and Sustainable Technology, Solar Household Energy, Inc., Numerical Logics, Inc., and the Atmospheric Sciences Research Center at The University at Albany.

Wild, Matthew, L. Wind Drives Growing Use of Batteries, New York Times, July 28, 2010, pp. B1.

Wikia 2011, Solar cookers world network [Online]. Available from http://solarcooking. wikia.com/wiki/Solar Cookers_World_Network_%28Home%29

Worldwatch Institute (January 2012). "Use and Capacity of Global Hydropower Increases"

"World Wind Energy Report 2010" (PDF). Report. World Wind Energy Association. February 2011, http://www.worldwatch.org/node/9527

Zweibel, K., Should solar photovoltaics be deployed sooner because of long operating life at low, predictable cost? Energy Policy (2010), doi:10.1016/j.enpol.2010.07.040

INDEX

Note: Page numbers with "f" denote figures; "t" tables; "b" boxes.

COLOR PLATE

Combined Power Plant

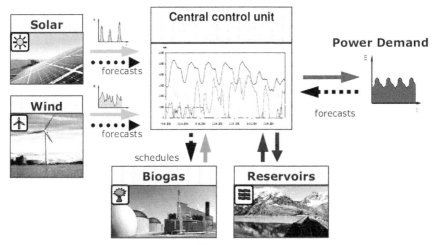

Plate 1

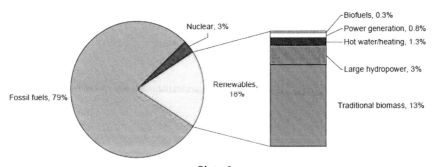

Plate 2

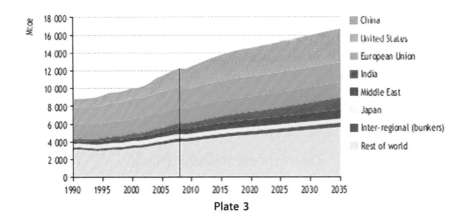

Plate 3

World's Liquid Fuels Supply

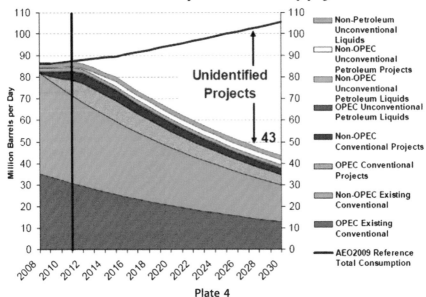

Plate 4

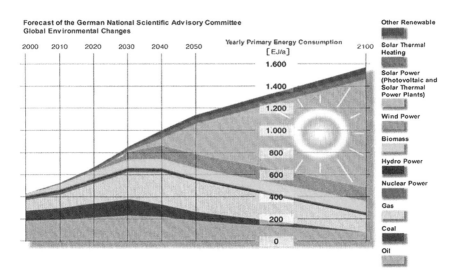

Plate 5

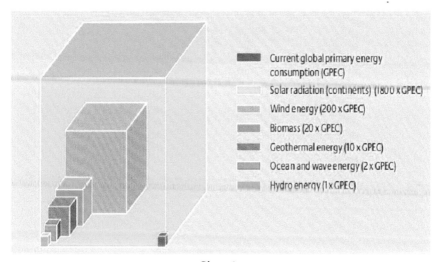

Plate 6

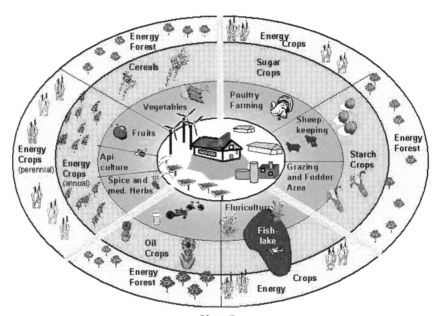

Plate 7

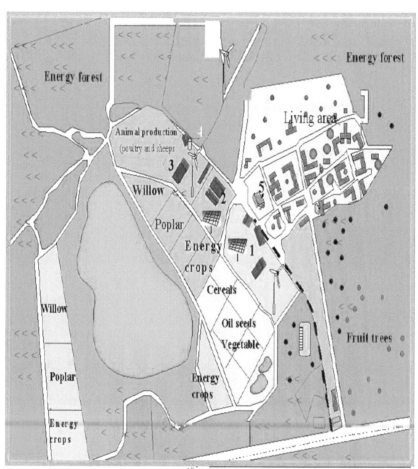

1 Thermal and Power unit (Biomass, Wind, Solar) 2 Pelleting, Oil mill, Ethanol unit
3 Animal husbandry 4 Biogas unit 5 Administration

Plate 8

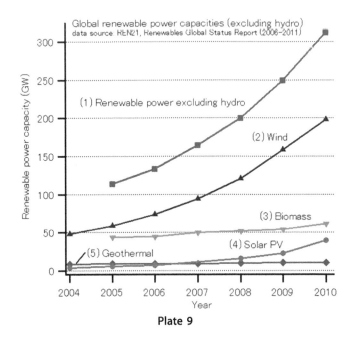

Plate 9

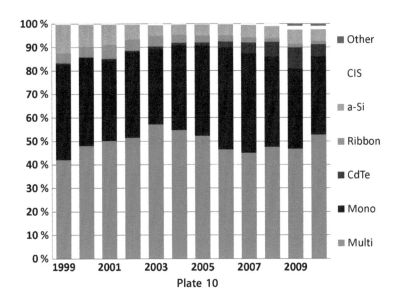

Plate 10

(a)

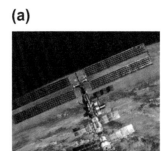

Plate 11

(b)

Plate 12

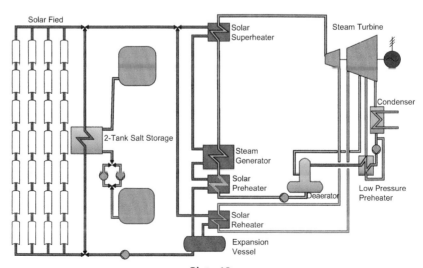

Plate 13

Plate 14

Plate 15

Plate 16

(A)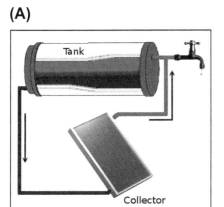

(B)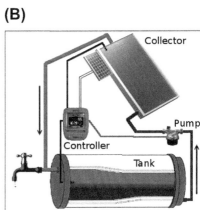

Plate 17

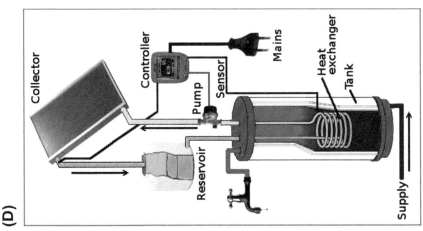

(C)

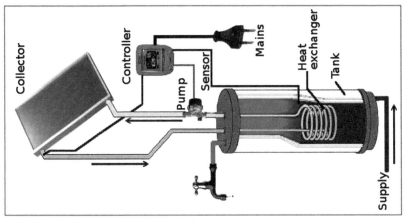

(D)

Plate 18

Plate 19

Plate 20

Plate 21

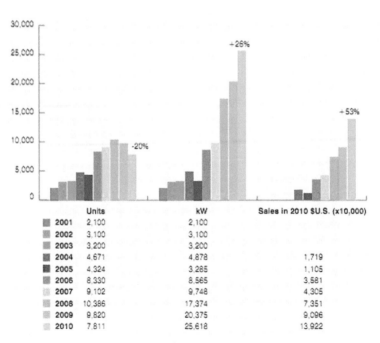

Plate 22a

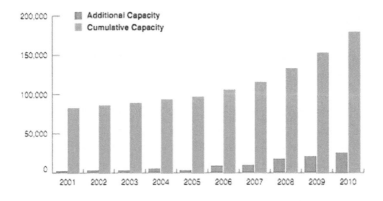

Plate 22b

Plate 23

Plate 24

Plate 25

(a)

Plate 26

Plate 27

(a)   **(b)**

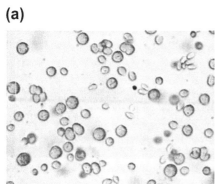

Plate 28

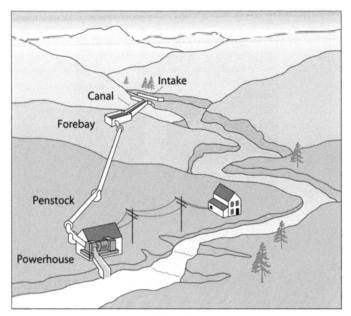

Plate 29

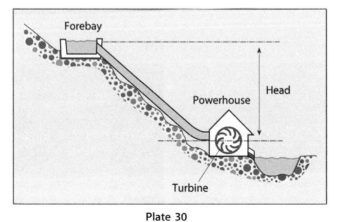

Plate 30

Plate 31

Plate 32

Plate 33

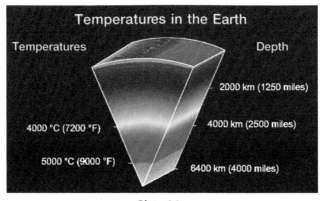

Plate 34

Plate 35

Plate 36

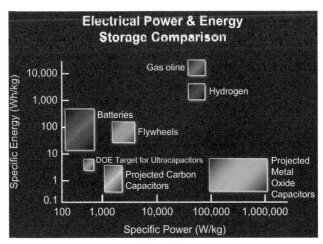

Plate 37

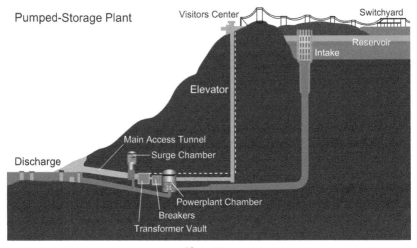

Plate 38

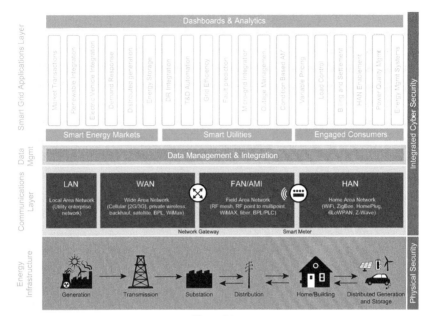

Plate 39

Plate 40

Plate 41

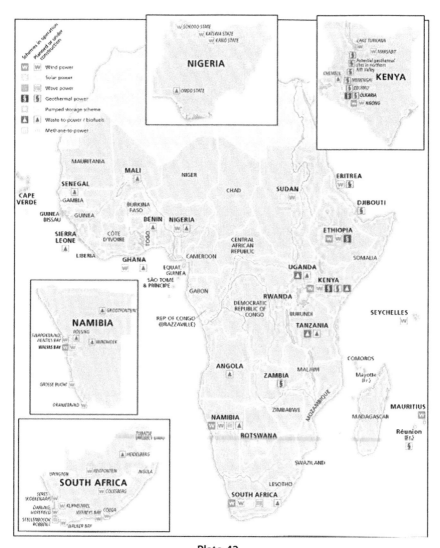

Plate 42

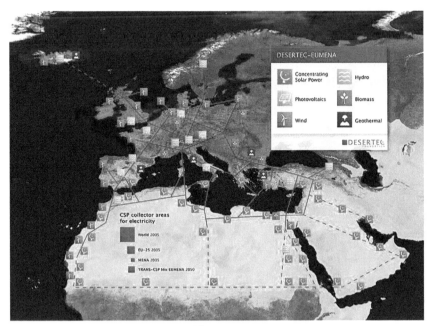

Plate 43

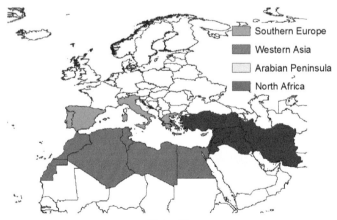

Plate 44

Plate 45

Plate 46

Plate 47

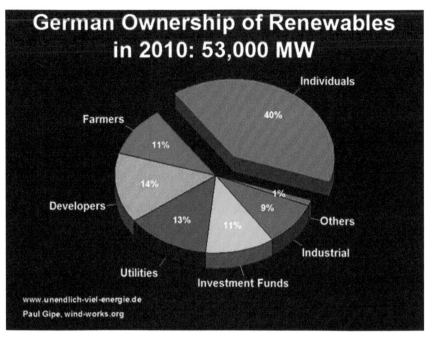

Plate 48

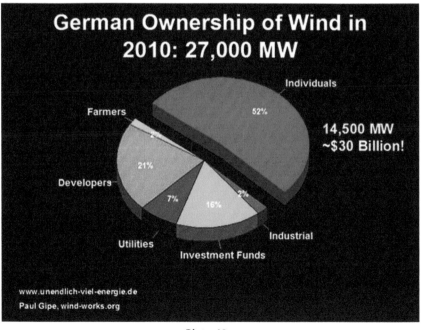

Plate 49

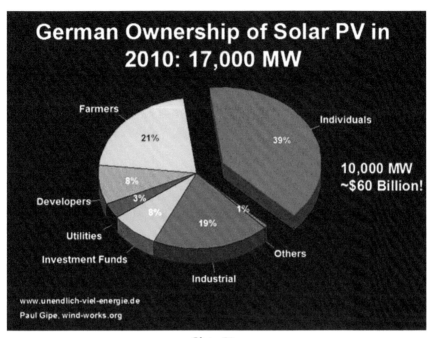

Plate 50

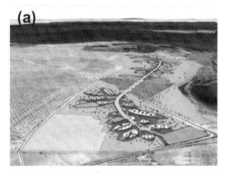

Plate 51

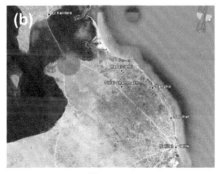

Plate 52

Plate 53

Plate 54

Plate 55

Plate 56

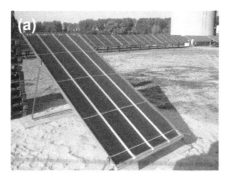

Plate 57

Plate 58

Plate 59

Plate 60

Plate 61

Plate 62

Plate 63

Plate 64

Printed and bound by CPI Group (UK) Ltd, Croydon, CR0 4YY

03/10/2024

01040415-0011